Into the
ELECTRIC AGE

1880 TO 1900

Published by The Reader's Digest Association, Inc.
London • New York • Sydney • Montreal

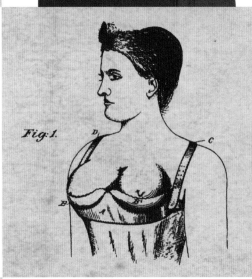

Fig:1.

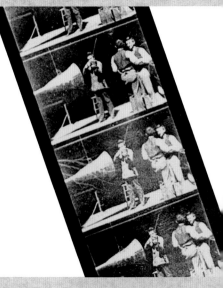

RUECK & TEXTOR PROPRIÉTAIRES

HOTEL DE LA GRANDE BRETAGNE NICE

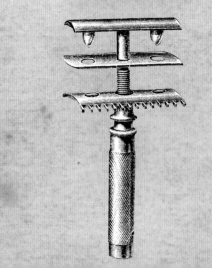

Contents

Introduction

The 19th century drew to a close in subdued colours. Democracy had made progress in Europe, but it was also threatened by anarchist outrages and by anti-Semitism, notably manifested in France in the Dreyfus affair. From 1876 on, for the first time since the Industrial Revolution had got properly under way, the world's economy had been slowing, with financial crises and agricultural downturns affecting both Europe and the USA. Population growth slowed, immigration to the New World increased and colonial empires continued to expand, driven as much by the hunt for raw materials and fresh outlets for exports as by any claimed 'civilising' mission.

It was in the USA – at this time still an emerging power – that electricity, the vital new source of energy, established itself. Tamed, domesticated and democratised, its unseen presence would soon permeate the hearts of cities thanks to the discovery of alternating current and the development of the first power stations, many of which exploited the energy of rivers.

The economic recession, limited as it was in scope, did not hamper scientific progress. Researchers penetrated the secrets of matter and of the human body. Much that had formerly been invisible now came to light, from x-rays to radioactivity from neurons to viruses, as scientists threw themselves wholeheartedly into their work, sometimes at the risk of their own lives. Yet the intellectual era of scientific positivism, when it was believed that

science would answer all of the world's problems, was coming to an end.

The creative imagination reasserted itself in reaction to the prevailing mood of materialism and economic gloom. In the arts realism and naturalism gave way to symbolism and Impressionism. Literature saw the first works of science fiction, while the coming of cinema gave the public a whole new way of escaping the realities of life. Some sought new heroes in the revived Olympic Games, reflecting a spirit of idealism that crossed frontiers.

The motor car made its first appearance, playing to the restlessness of the age. At the time few suspected that cars would be the driving force of a new phase of economic growth, one powering the sunny years before the First World War that would become known by its French name, the Belle Epoque.

The editors

Reaching for the sky
The reborn city of Chicago gave architects a green light to create a city for the 20th century. They and their buildings rose to the challenge on the shores of Lake Michigan. Harnessing new materials and techniques, Chicago pioneered the skyscraper, producing the archetypal city skyline.

◄ The closing decades of the 19th century saw electricity become part of daily life, as it brightened homes and workplaces, helping to reduce the incidence of industrial accidents

▼ The Machine Hall at the Paris World's Fair of 1889 was a monument to toughened industrial glass, the development of which made possible the glass-and-steel architecture of the late 19th century possible

◄ The electric light bulb played a central role in the adoption of electricity by the public; the bulb shown here (left) featured in an advertisement for the Westinghouse Electric Company

Social historians have tended to see the last two decades of the 1800s as the *fin de siècle*, a period of endings rather than beginnings. Yet for all the contradictions that marked the era, it nonetheless nurtured seeds that would blossom in the

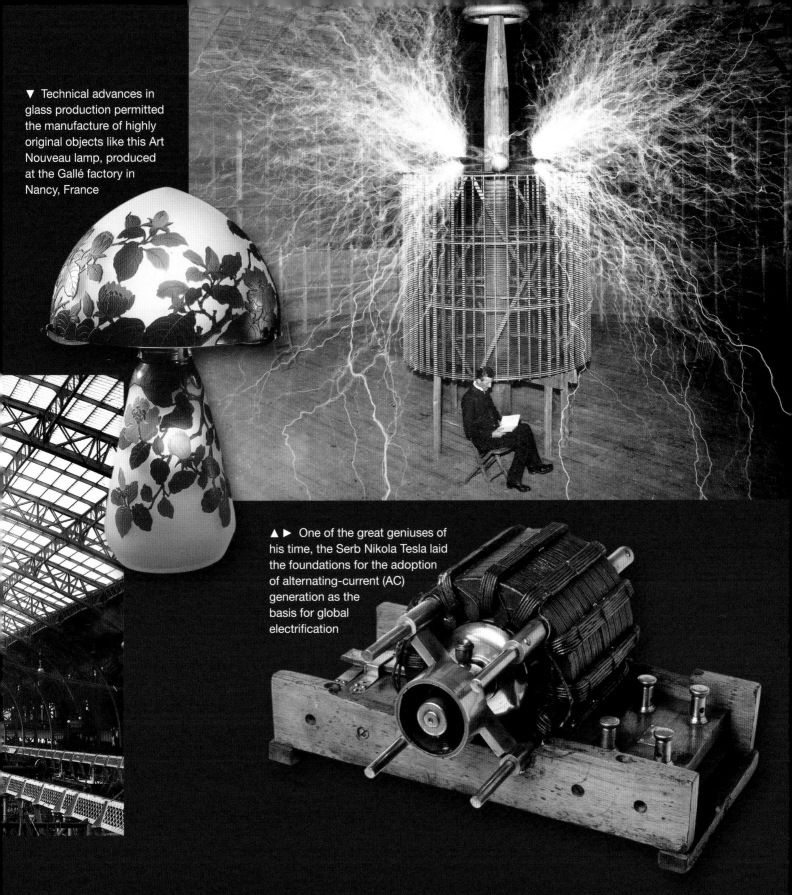

▼ Technical advances in glass production permitted the manufacture of highly original objects like this Art Nouveau lamp, produced at the Gallé factory in Nancy, France

▲ ▶ One of the great geniuses of his time, the Serb Nikola Tesla laid the foundations for the adoption of alternating-current (AC) generation as the basis for global electrification

20th century. It was in 1895 that Nikola Tesla developed alternating current (AC) electricity generation, paving the way for global electrification and for a second industrial revolution that got under way in the years after 1900. Electric ovens and kettles duly put in

◄ Invented in successive years, in 1886 and 1887, Linotype and Monotype machines revolutionised the composition of text in the printing industry, enabling typesetters to deal with entire lines and blocks of type rather than individual characters

◄ Invented almost by accident in 1885, motorcycles were initially playthings for the rich

an appearance; so too, sadly, did the electric chair. Meanwhile, another revolution was in the making in the field of transport. Bicycles were made more comfortable with the invention of inflatable rubber tyres. Equipped with motors from 1885 on, they

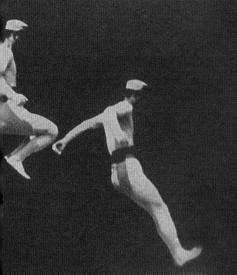

▼ Invented by John Boyd Dunlop in 1888, inflatable rubber tyres made bicycling a more comfortable and enjoyable experience; Edouard Michelin (below) introduced the removable tyre a year later

▲ To analyse and measure human and animal movement, Étienne Jules Marey came up with a way of recording figures in motion, in the process contributing a big step towards cinematography

◄ Taking inspiration from the photographic revolver devised by French astronomer Jules Janssen, in 1882 Marey mounted a camera on a rifle butt to study the flight of birds; his device enabled him to take a dozen shots a second

► The first Kodak camera, produced in 1888, was equipped with a celluloid film that could hold 100 images and was, in the words of its inventor George Eastman, 'as easy to use as the pencil'

were transformed into motorcycles. The chief breakthrough was the automobile, which introduced a new era of liberty to the world of travel. Born of the work of visionaries like Carl Benz and Gottlieb Daimler, the machines powered by the combustion

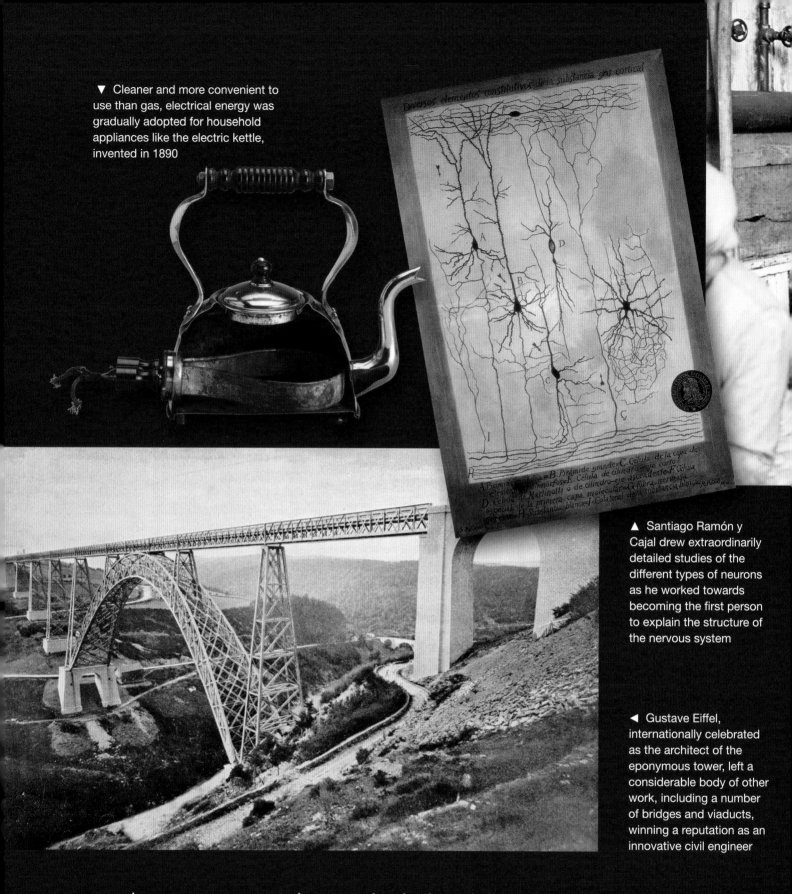

▼ Cleaner and more convenient to use than gas, electrical energy was gradually adopted for household appliances like the electric kettle, invented in 1890

▲ Santiago Ramón y Cajal drew extraordinarily detailed studies of the different types of neurons as he worked towards becoming the first person to explain the structure of the nervous system

◀ Gustave Eiffel, internationally celebrated as the architect of the eponymous tower, left a considerable body of other work, including a number of bridges and viaducts, winning a reputation as an innovative civil engineer

engine were soon taken up by industrialists, among them Charles Rolls and Armand Peugeot as well as the pioneer of mass production, Henry Ford. Something of the excitement of the times found its way into the novels of Jules Verne, whose science-fiction

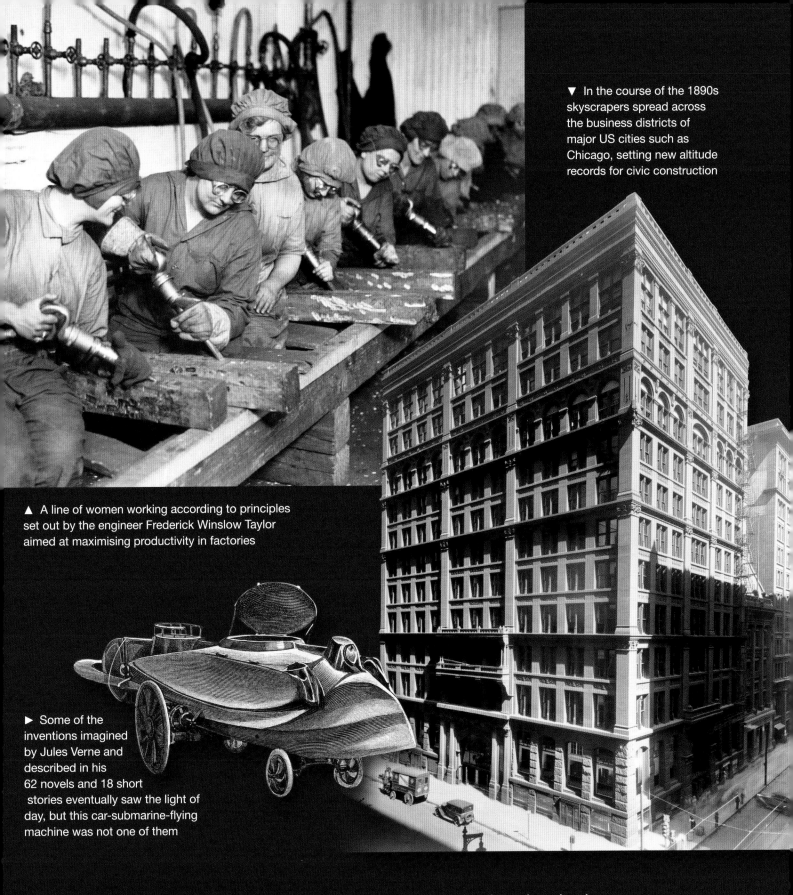

▼ In the course of the 1890s skyscrapers spread across the business districts of major US cities such as Chicago, setting new altitude records for civic construction

▲ A line of women working according to principles set out by the engineer Frederick Winslow Taylor aimed at maximising productivity in factories

► Some of the inventions imagined by Jules Verne and described in his 62 novels and 18 short stories eventually saw the light of day, but this car-submarine-flying machine was not one of them

fantasies matched the spirit of the age, capturing both the enthusiasm stirred by recent discoveries and the anticipation of ones still to be made. One dream that did became a reality at this time was the photographic analysis of bodies in motion,

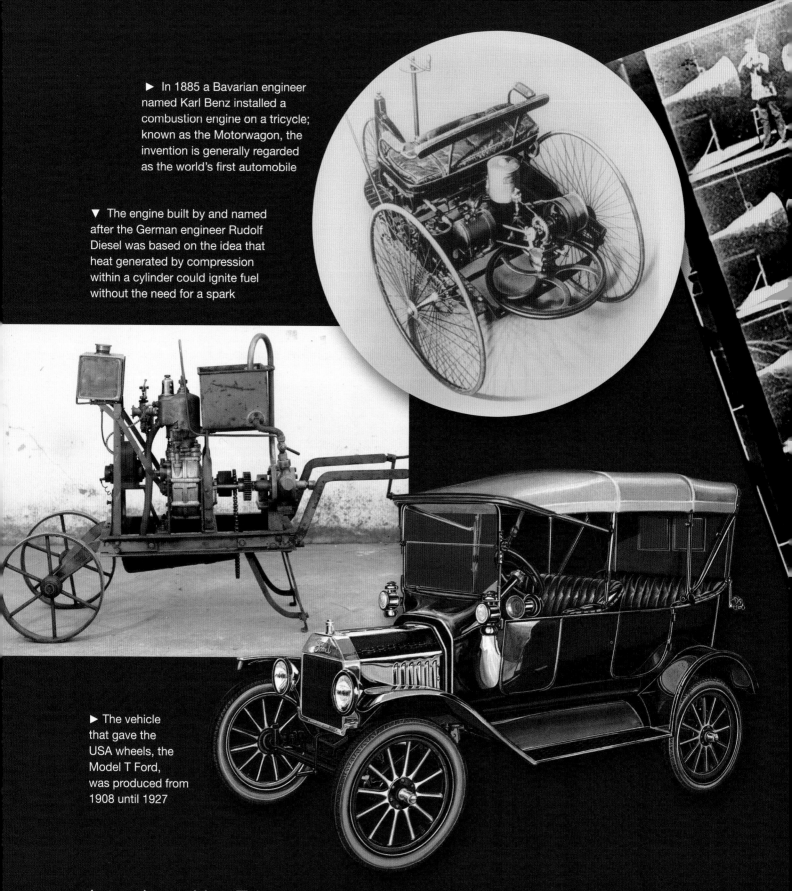

► In 1885 a Bavarian engineer named Karl Benz installed a combustion engine on a tricycle; known as the Motorwagon, the invention is generally regarded as the world's first automobile

▼ The engine built by and named after the German engineer Rudolf Diesel was based on the idea that heat generated by compression within a cylinder could ignite fuel without the need for a spark

► The vehicle that gave the USA wheels, the Model T Ford, was produced from 1908 until 1927

introduced by Etienne Jules Marey, which proved to be the immediate precursor of the cinema. By 1900 the Lumière brothers were entrancing Paris audiences with documentary footage of workers pouring out of factories, while Georges Méliès was already

◀ The brothers Antoine and Louis Lumière made improvements to Edison's Kinetoscope to invent the Cinematograph, patented in February 1895

▶ Pioneer film-maker Georges Méliès created a fantasy world that fascinates viewers to this day through the ingenuity of his special effects; his future wife Jehanne d'Alcy is seen here in his 1899 version of *Cinderella*

▼ Slapstick and disasters were two staples of the silent cinema, which required actors to express powerful emotions without the aid of speech

at work on the first fantasy films. 'Faster, higher, stronger' was the motto of the Olympic Games, revived from ancient Greek tradition in 1896, but it could also have applied to contemporary trends in architecture. From Chicago skyscrapers to the Eiffel Tower in Paris,

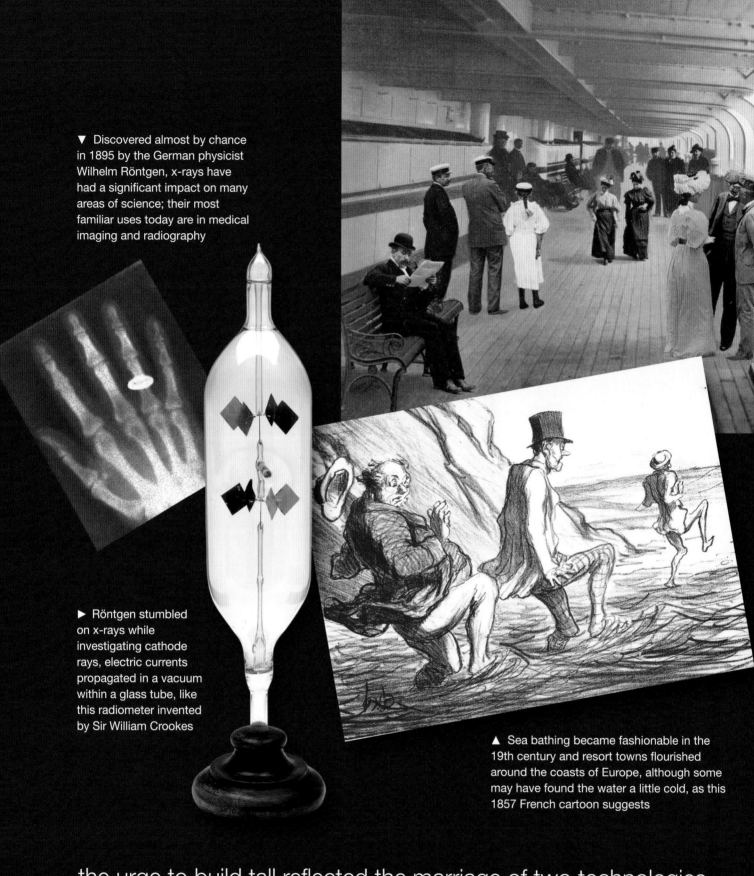

▼ Discovered almost by chance in 1895 by the German physicist Wilhelm Röntgen, x-rays have had a significant impact on many areas of science; their most familiar uses today are in medical imaging and radiography

▶ Röntgen stumbled on x-rays while investigating cathode rays, electric currents propagated in a vacuum within a glass tube, like this radiometer invented by Sir William Crookes

▲ Sea bathing became fashionable in the 19th century and resort towns flourished around the coasts of Europe, although some may have found the water a little cold, as this 1857 French cartoon suggests

the urge to build tall reflected the marriage of two technologies – mechanised construction methods and new techniques for strengthening steel. Medical research continued to make spectacular progress – in the neuron theory championed by

◄ The development of affordable transport widened the horizons of passengers like these on board the *Kaiserin Maria Theresa* in 1895, who could cross the seas in the relative comfort of an ocean liner

▲ Using this instrument of his own devising, in 1896 Joseph John Thomson made the revolutionary discovery that atoms can emit sub-particles of matter – electrons – thus putting paid to the ancient belief that atoms are indivisible

▼ ► Competitors line up for the 100m sprint at the first modern Olympic games, held in Athens in 1896; race times were recorded using Longines watches

Santiago Ramón y Cajal, in the discovery of viruses and blood groups, and in the identification of mosquitoes responsible for spreading malaria. Wilhelm Röntgen developed x-rays, permitting doctors to 'see' inside the body. Even more fundamentally,

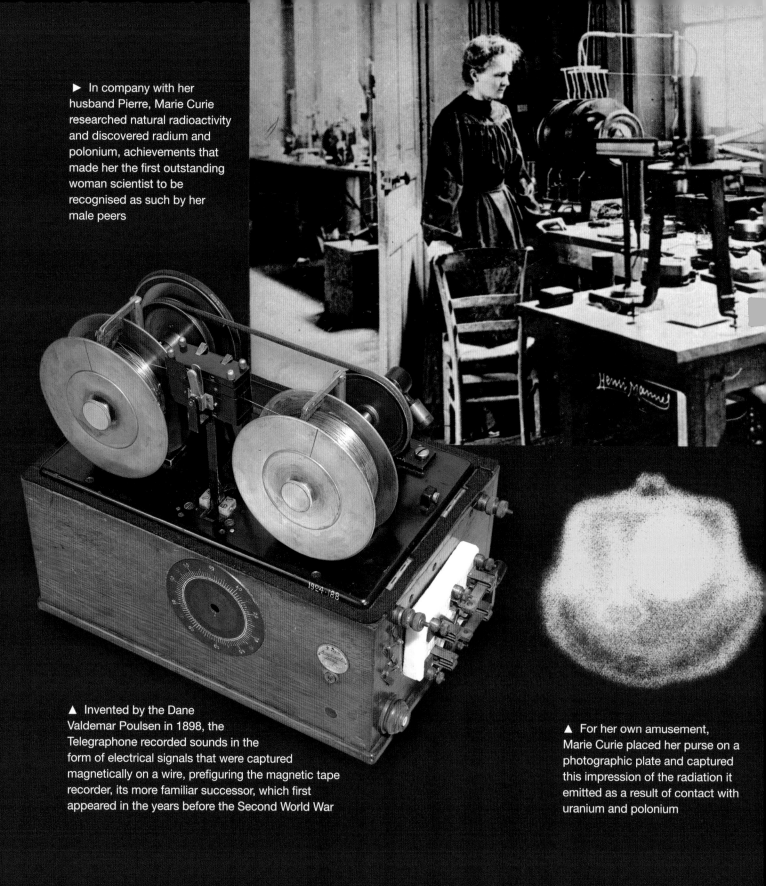

► In company with her husband Pierre, Marie Curie researched natural radioactivity and discovered radium and polonium, achievements that made her the first outstanding woman scientist to be recognised as such by her male peers

▲ Invented by the Dane Valdemar Poulsen in 1898, the Telegraphone recorded sounds in the form of electrical signals that were captured magnetically on a wire, prefiguring the magnetic tape recorder, its more familiar successor, which first appeared in the years before the Second World War

▲ For her own amusement, Marie Curie placed her purse on a photographic plate and captured this impression of the radiation it emitted as a result of contact with uranium and polonium

research into the nature of matter revolutionised physics. Joseph John Thomson demonstrated that atoms are composed of a nucleus surrounded by electrons – and could therefore be split. At approximately the same time, Henri Becquerel was revealing the

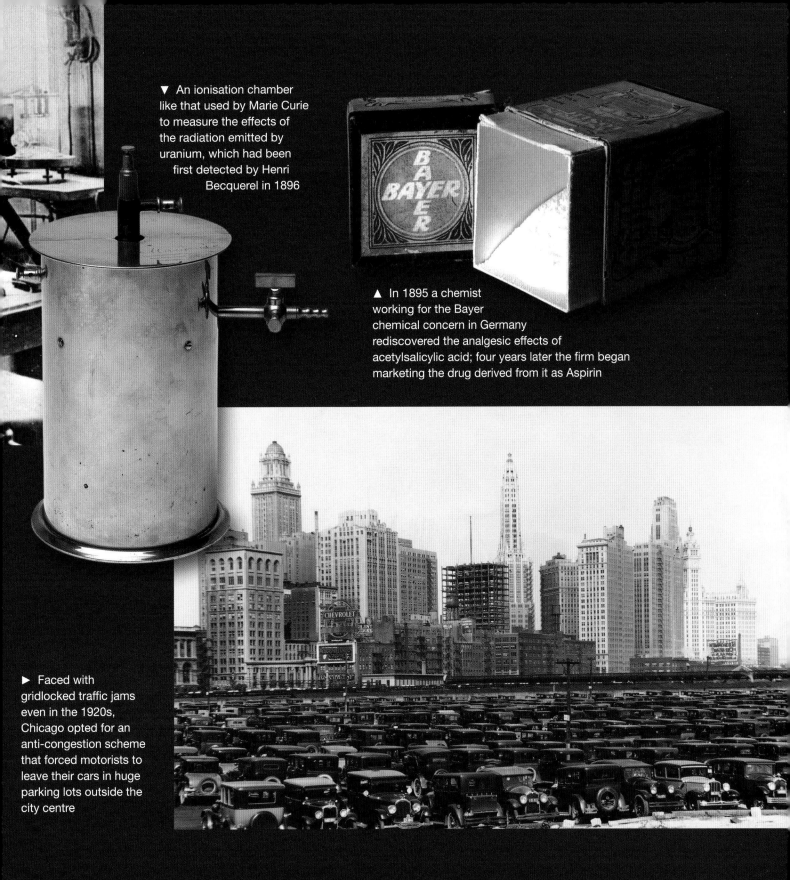

▼ An ionisation chamber like that used by Marie Curie to measure the effects of the radiation emitted by uranium, which had been first detected by Henri Becquerel in 1896

▲ In 1895 a chemist working for the Bayer chemical concern in Germany rediscovered the analgesic effects of acetylsalicylic acid; four years later the firm began marketing the drug derived from it as Aspirin

► Faced with gridlocked traffic jams even in the 1920s, Chicago opted for an anti-congestion scheme that forced motorists to leave their cars in huge parking lots outside the city centre

presence of natural radioactivity, which also attracted the attention of Marie and Pierre Curie. In many respects, then, the closing years of the 19th century were less an epilogue for the Victorian age than a foretaste of things to come.

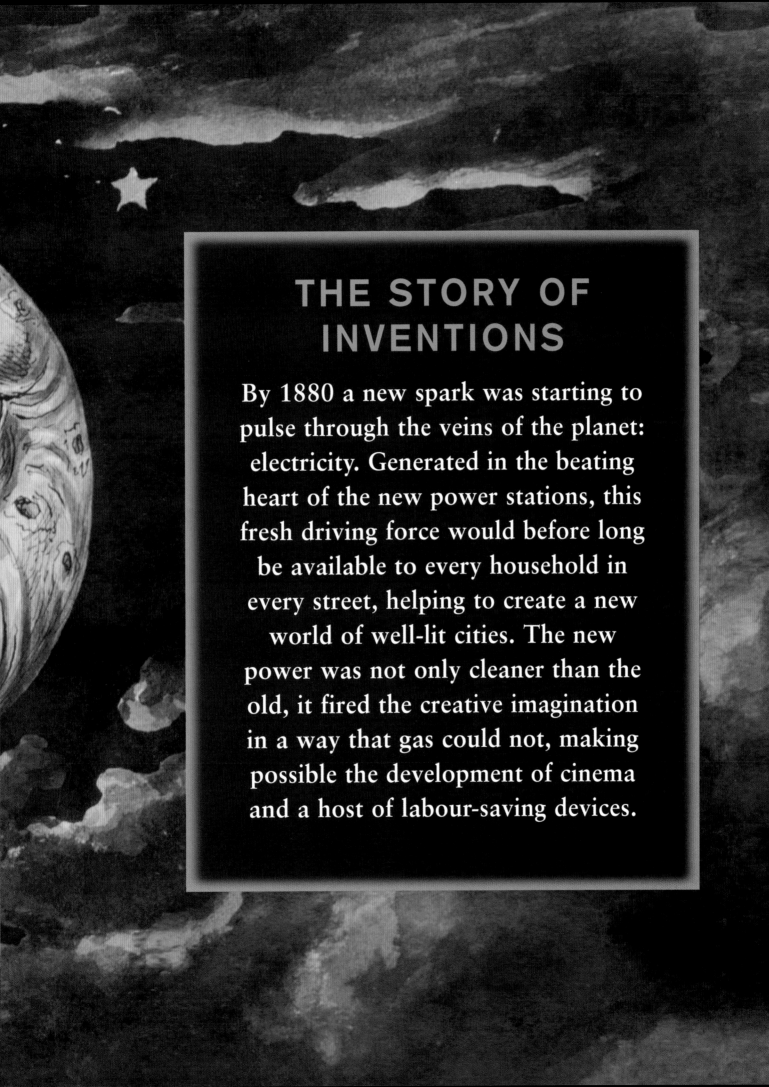

THE STORY OF INVENTIONS

By 1880 a new spark was starting to pulse through the veins of the planet: electricity. Generated in the beating heart of the new power stations, this fresh driving force would before long be available to every household in every street, helping to create a new world of well-lit cities. The new power was not only cleaner than the old, it fired the creative imagination in a way that gas could not, making possible the development of cinema and a host of labour-saving devices.

POWER STATIONS – 1881
Electric light for everyone

The establishment of power stations and supply networks made it possible first for selected urban districts and then for entire cities to be connected to mains electricity. In their wake new industries sprang up and electric lighting brightened the streets.

City illumination
Electricity lights up a New York street in 1889 (right). When the city's first power station came on line, only 52 households expressed an interest in signing up for the service.

Tapping into Nature's might
A power station on the Niagara River (below) in 1899. With a flow of more than 2,800m³/s, the falls have served as a source of energy since the mid 18th century.

In 1881 the first power station worthy of the name was built at Godalming in Surrey. It consisted of an AC alternator and dynamo, manufactured by the Siemens company in Germany, powered by the twin wheels of a local water mill. It was not very efficient and ceased operation after just three years for lack of public support. By that time, others had come into service, notably the coal-powered Pearl Street plant in New York, built by Thomas Edison's company in 1882.

With the dawning of easily available electricity, the days of gas lighting, which had proved expensive, evil-smelling and dangerous, were numbered. Edison aimed to sell electricity from his power stations in a ready-to-use form. Up to that time anyone who wanted to take advantage of the potential benefits of electricity had to provide their own energy source in the form of a dynamo or some other type of generator. This had limited its use to leading industrialists, big institutions and the super-rich. Edison managed to find solutions for the technical problems involved in the distribution of electricity to the public. Key to his success was the development of a practical light bulb that could bring electric light into any home. Edison's electricity was still expensive, but it was within the means of many more people and soon coal-powered generating plants along Pearl Street lines were spreading across the USA and Europe.

THE DAM BUILDERS

In 1879 an American engineer named Lester Pelton invented a hydroelectric turbine capable of efficiently capturing the energy of falling water and transferring it to an alternator. His discovery set in motion a drive to harness rivers and waterfalls by building dams. In 1895 a power station was constructed to tap into the might of the Niagara Falls, and by the following year a dam on the St Lawrence River was supplying electricity to Montreal in Canada. By 1920 hydroelectric production was providing 40 per cent of the USA's electricity. Europe followed suit; in France alone, 51 dams were built between 1920 and 1940. Up to the 1960s, most of the power passing through the world's cables was generated from hydroelectric sources.

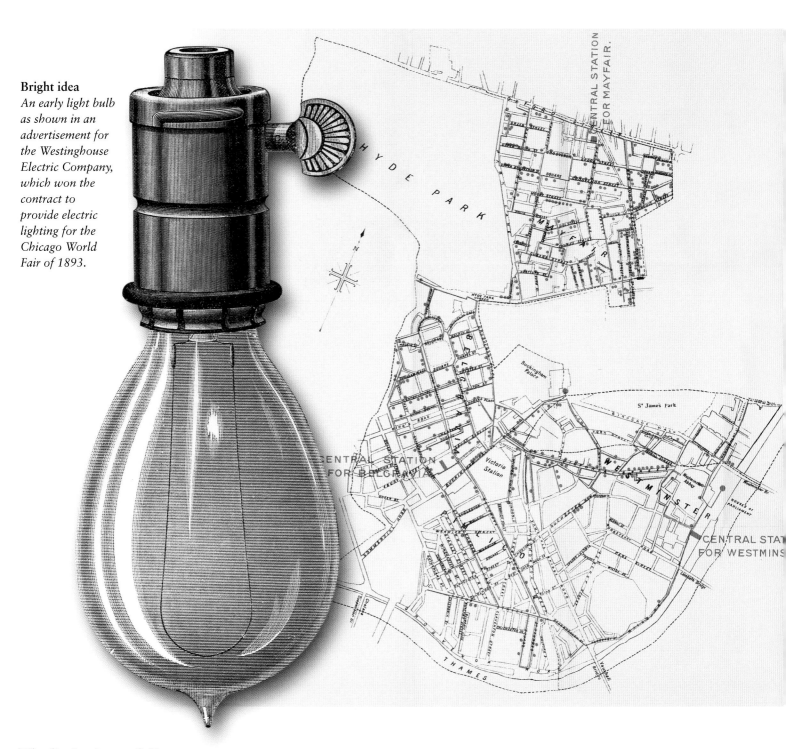

Bright idea
An early light bulb as shown in an advertisement for the Westinghouse Electric Company, which won the contract to provide electric lighting for the Chicago World Fair of 1893.

The limitations of direct current

Yet the Pearl Street plant and its imitators had a flaw that threatened to impede the adoption of electricity as a major source of energy around the world: they produced power and supplied their customers with direct current (DC) at a relatively weak tension of 110 volts. The system was incapable of producing and transporting higher voltages over long distances. At best, Edison could only construct small, local generating plants supplying the needs of districts a few square miles in size. Something more ambitious was needed to provide for a city and the growing demands of

industry, let alone an entire nation. Under such conditions, the economic viability of electricity as an energy source was far from assured.

AC on a roll

In 1881 two inventors – John Dixon Gibbs from England and a Frenchman named Lucien Gaulard – made a discovery that would radically change the situation. They developed a transformer capable of increasing the tension of a current to previously unattainable levels. The power of a current reflects both its voltage and the strength of the electrical charge as measured in amps, but power loss is entirely a

Localised supply
Dating from 1891, the plan above shows generating stations run by the Westminster Electric Supply Corporation in the Mayfair, Belgravia and Westminster districts of London.

21

ALTERNATORS

Like dynamos, alternators convert mechanical into electrical energy. But where dynamos produce a continuous current, alternators – as the name suggests – generate an alternating current in which the electric charge repeatedly changes direction. The mechanism consists of a magnet; a rotor, which rotates within a fixed coil of copper wire; and a stator. The rotor's motion causes alterations in the magnetic field surrounding the stator, thereby creating through the physical principle of induction a current whose direction alternates in accordance with the movements of the magnet.

Alternators at work *A series of small machines generating alternating current in a Westinghouse power station in New York in the late 1880s. As the holder of the requisite patents, the Westinghouse company was eventually able to impose the AC standard that it championed across the USA.*

function of the latter. To transport high levels of power over long distances with minimal loss, the best solution is to produce a relatively weak current at a very high voltage.

Understanding this principle, the American entrepreneur George Westinghouse bought up Gaulard and Gibbs' patents in 1885. And since transformers only operate on alternating current (AC) he hired its champion, Nikola Tesla, a former Edison employee, to build AC power stations. That same year Westinghouse built a small, steam-powered plant in Great Barrington, Massachusetts, that produced enough electricity to light the town at a tension of 500 volts. For transportation the voltage was increased to 3000V, then lowered to just 100V to power light bulbs. Within the next year, 30 similar stations were constructed across the USA. Success was in sight. From that time on it was possible to supply wide areas from a limited number of power plants.

The advent of hydroelectricity

The development of AC made it possible to exploit the hydroelectric potential of rivers in mountainous regions far from cities. Dams were built and water became as important as coal to the production of electricity. Improved methods of transporting current also allowed supply networks to be linked together. Eventually, after the major power suppliers reached agreement on industry standards, the networks became first regional, then national,

Power pioneers
Emil Rathenau (below, on the left) and Thomas Edison pose in front of a turbine in Berlin in 1911. The German industrialist had bought patent rights from Edison to found the AEG company 28 years earlier.

and eventually international in their reach. The amount of electricity supplied by the system also grew along with the voltage carried in the lines, which rose from 22,000V in 1896 to 100,000V in 1908, reaching 220,000V by 1921 and 1 million volts in more recent years .

Electricity's triumph

The last years of the 19th century prepared the way for electricity's universal adoption. The development of powerful electric motors put the new energy source at the disposal not just

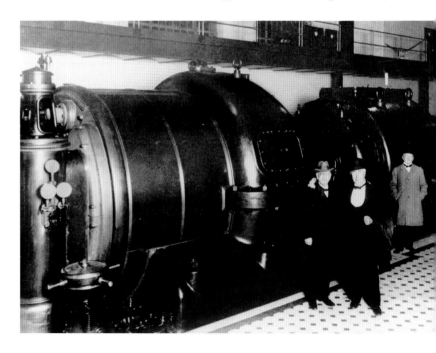

of large-scale industries, as steam had been, but also of small tradesmen. The first of the electrical appliances that would make life more comfortable in the course of the 20th century were already coming on the market.

Above all, it was electric lighting that won over the public. A new life-style dawned for people who took it up, as their days became artificially longer and leisure hours expanded. Improved lighting in workshops and factories reduced accidents and increased productivity. New industries sprang up to take advantage of electrochemical and electrometallurgical techniques. Aluminium production spiralled, for example, when electrical methods replaced chemical ones in the treatment of bauxite. New forms of transport also emerged: first electric trams and trains, and then electrically powered underground railways.

Even so, the spread of electrification was relatively slow. The USA led the world in the production of electricity, but in 1910 just 15 per cent of American homes had a mains connection; the equivalent figure for Britain was 2 per cent and even by 1936 only 12,000 homes in Britain were connected to the National Grid (although there were still dozens of local generating stations). Conveniences like electric refrigerators and washing machines only became standard household items from 1930 on in the USA, while in much of Western Europe they would not arrive until the 1950s. It is worth remembering that even today almost a quarter of the world's population lives without electricity.

THE TRANSFORMER

These machines are used either to increase or lower the voltage of a current. In both cases the principle at work remains the same. The current to be transformed passes through one of two copper induction coils positioned alongside one another. The alternating current in the first of these generates a changing magnetic field that, following the principle of induction, creates a different current in the second coil. Assuming that this one has been properly set up – for example with respect to the diameter of the copper wire and the number of coils employed – the current's voltage can be modified at will.

Transformers at work
In power stations transformers like these (right) play a vital role in modulating the voltage of currents.

THE JOULE EFFECT

Electric current is caused by the displacement of electrons inside conductive materials like metals. Yet the electrons have only limited freedom of movement. Most materials present a resistance that causes them to lose energy, which is dissipated in the form of heat – the reason why cables grow warm when current passes through them. The energy thus produced is proportional to the square of the strength of the current.

Stereophonic sound 1881

Entertainment from afar
Guests listen to electrophone transmissions in the salon of a Paris hotel in 1892.

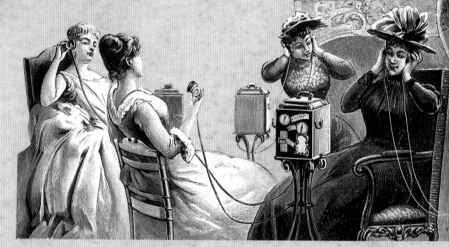

FROM THE ELECTROPHONE TO HOME CINEMA

Multitrack recording, which allows different sound sources to be recorded separately, was introduced to the cinema as part of the Cinerama package developed by Fred Waller of Paramount Pictures in 1952. It was also a feature of CinemaScope, a rival system premiered the following year. In 1982 Ray Dolby introduced the Dolby Surround system, featuring up to 16 audio tracks. The advent of digital sound further expanded the domestic audience for multitrack with the coming of home cinema in the first decade of the 21st century.

Best known as a pioneer of aviation, France's Clément Ader also made breakthroughs in the field of telephony. One of his most significant discoveries was stereophonic sound. For the Paris World's Fair of 1881 he developed a system capable of transmitting live performances at the Opera to an exhibition hall some 2km away, via lines laid through the sewers. Listeners were equipped with twin earpieces fed by two separate telephone lines, one capturing sound from the right and the other from the left-hand side of the stage. In Britain this two-channel process became known as the electrophone, and it continued in use for several decades until it was killed off by the spread of the gramophone (invented in 1878) and the advent of radio (after 1897). Stereo sound would not be adopted for records until 1958 and for radio transmission until 1961.

The trolley bus 1882

From the early 19th century, public transport came to play an increasingly essential role in modern life, but trams, which ran on rails, required a huge infrastructure investment. To reduce this, the German industrialist Werner von Siemens invented an alternative vehicle that he called the *Elektromote*. It was powered by an electric current from overhead wires, fed to the vehicle down a pole attached to a 'trolley' that ran along the wires. A first public demonstration took place in Berlin on 29 April, 1882.

The invention was a huge success and Siemens went on to supply customers around the world. At one time there were 50 separate trolley-bus systems in the UK alone. But eventually trolley buses, which were restricted to the routes of the overhead wires, fell out of favour, to be replaced by more flexible petrol-driven buses and private cars; the last British service, in Bradford, was withdrawn in 1972. Interest has revived in recent years because trolleys are more environmentally friendly than their oil-fuelled rivals, and services have been maintained or restored in some American and European cities.

The *Elektromote* Siemens is at the wheel (below) for this trial run in the Berlin suburb of Halensee.

The capillary-feed pen 1884

The year was 1883 and an American insurance broker named Lewis Edison Waterman was about to close one of the biggest deals of his career. With a flourish he produced a stylish fountain pen he had recently acquired, but then, to his horror, the implement malfunctioned, spewing ink all over the document. Before Waterman could draw up a fresh contract, his dissatisfied client had signed up with a rival firm.

The pen had cost him the deal, and so Waterman set about devising a more reliable device which would take advantage of a simple principle of physics – the law of capillarity, by which liquids are drawn to small spaces. He set about replacing the single feeding channel from the ink reservoir of the rejected pen with multiple grooves. These served to regulate the ink flow and led him to call his new improved pen the Regular.

Waterman patented the pen in 1884 and was confident enough of its reliability to sell it with a five-year guarantee. The capillary-feed fountain pen increased in popularity from 1891 on with the invention of replaceable cartridges. It only began to fall out of favour after the introduction of ballpoint pens in 1938 and then of felt-tips in 1962.

In the service of peace *An advertisement from 1919 (above right) presents Waterman pens as an 'arm of peace'; one had been used to sign the Treaty of Versailles after the end of the First World War.*

Porte Plume "Ideal"
Waterman
l'Arme de la paix

TRAITÉ de PAIX

1919

FROM GOOSE QUILLS TO STEEL NIBS

Metal-tipped pens have been known since antiquity, although the most common writing implement at the time was the quill, fashioned from the feathers of geese, crows, ducks or vultures. At the start of the 19th century, when paper quality improved and steel became widely available, metal pens entered the industrial age. In 1828 a Birmingham industrialist named Joseph Gillott invented the first machine designed to mass-produce them. He found a ready market for his product thanks to the growing requirements of civil servants and of teachers and their pupils. By 1840 Gillott's factory was using 50 tonnes of steel a year to produce some 75 million nibs. In the later years of the 19th century his products and those of his rivals became part of everyone's childhood, used to practice upstrokes and downstrokes in thousands of primary schools across the land.

The sharp end of literacy *Steel pen nibs like these French-made examples (left) provided a regular ink flow that permitted generations of schoolchildren to master the art of joined-up writing.*

25

THE DUSTBIN – 1884
Putting garbage in its place

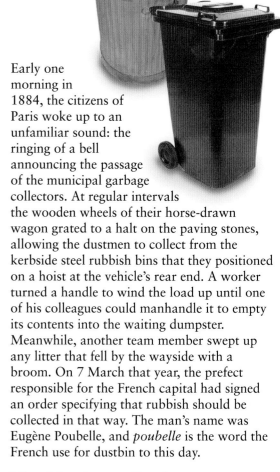

Until the late 19th century the collection of urban waste was generally left to farmers in search of manure or pig food and rag merchants who recycled anything that was saleable. But a growing awareness of the part played by microbes in spreading epidemic diseases eventually persuaded the civic authorities in Europe's great cities to take drastic public hygiene measures.

Early one morning in 1884, the citizens of Paris woke up to an unfamiliar sound: the ringing of a bell announcing the passage of the municipal garbage collectors. At regular intervals the wooden wheels of their horse-drawn wagon grated to a halt on the paving stones, allowing the dustmen to collect from the kerbside steel rubbish bins that they positioned on a hoist at the vehicle's rear end. A worker turned a handle to wind the load up until one of his colleagues could manhandle it to empty its contents into the waiting dumpster. Meanwhile, another team member swept up any litter that fell by the wayside with a broom. On 7 March that year, the prefect responsible for the French capital had signed an order specifying that rubbish should be collected in that way. The man's name was Eugène Poubelle, and *poubelle* is the word the French use for dustbin to this day.

The sickening smells of summer

It was not before time. The inhabitants of Paris had previously retained the medieval habit of emptying their slop buckets and food scraps out of the window into the gutter. In theory householders had a responsibility to clear the roadway in front of their homes, but few paid much attention, and the authorities proved unable to enforce the rules. Instead, people relied on dogs to scavenge organic waste and on rag merchants to sort through the debris that otherwise engulfed the thoroughfares.

Collecting the rubbish of Paris
An engraving from an illustrated magazine of 1884 shows a team of sanitary workers with a horse-drawn wagon clearing rubbish from a street on Paris's Left Bank.

SEWERS REPLACE CESSPITS

The task of preventing epidemics did not stop in the streets; it affected households, too. In earlier times houses and tenements had cesspits for human waste. Now, London led the way in replacing cesspits with sewers, gradually linking up waste disposal facilities. The Public Health Act passed in Britain in 1875 required all new-build housing to be equipped with internal drainage.

WHERE WASTE ENDS UP

Collecting rubbish is just one half of the process: the waste still has to be disposed of. Incineration first became an option in Britain with the introduction of the Destructor in 1874. This patented system burned mixed rubbish to produce steam that could be used to generate electricity. In all, 250 Destructors went into service before protests about the emission of ash and dust led to them falling out of favour in the 1920s. Landfill then became the popular disposal method; environmental concerns and a lack of suitable sites led to the introduction of a landfill tax in 1996, designed to encourage recycling.

Incineration plants today *Modern facilities like this one aim to consume waste as thoroughly as possible, but concerns remain about the emission of pollutants, particularly toxic dioxins.*

A similar situation in Britain had persuaded the authorities to take action nine years earlier. In 1875 the Public Health Act compelled municipal authorities to organise the collection of rubbish. The measure reinforced an earlier one that had first introduced the concept of waste management as a civic responsibility as far back as 1848. Many communities had been slow to take up the idea, though, and it was only in late Victorian times that dustbins became a feature on British streets. At the time more than 80 per cent of the waste put out was made up of ashes and cinders from coal fires – the 'dust' that gave dustbins their name.

The threat from microbes

The growing attention paid to waste disposal reflected a gradual acceptance of the microbial theory of disease. People had long suspected a link between urban squalor and the outbreak of epidemics, such as plague and cholera. John Snow, a London doctor, had helped to confirm the suspicion by identifying a water pump polluted with sewage as the cause of an outbreak of cholera in Soho in 1854. The actual microbes responsible were identified in 1861 and later Louis Pasteur's researches removed any remaining room for doubt.

In the cause of hygiene some municipal authorities took the lead in declaring war on germs. At the turn of the 20th century Prague, Vienna and Düsseldorf introduced covered

RIVER OF WASTE

In 2000, heavy rains caused the collapse of a garbage dump in the Philippine capital, Manila, killing 71 people.

dustbins that were removed when full and replaced with others that had been cleaned. In Britain the collected waste was sorted by hand, and salvageable materials such as glass or metals were recycled, while 'breeze' or hard-core cinders and ashes went to make 'breeze block' building materials.

Scavenging for a living
Children in many underdeveloped countries regularly scour open-air waste heaps in search of ironware, copper, aluminium, plastics and any other recyclable refuse that can be sold for a few pennies.

Glass gets tougher

Glass has been used since antiquity, not just for its transparency but also for its optical qualities and resistance to corrosion. Yet it was only in the final decades of the 19th century that heat-resistant and shatterproof varieties first made an appearance.

Gallery of glass
The Galerie des Machines *at the Paris World's Fair of 1889 (top right) was designed by architect Ferdinand Dutert and engineer Victor Contamin. At the time it was the largest metal-and-glass structure of its kind in the world, measuring 420m in length and 115m across and reaching a height of 43m.*

Precision instrument
Made of brass, this triple-lens Zeiss microscope (right) dates from 1870. Zeiss marketed his first, single-lens design in 1847; the earliest multi-lens model, the Stand I, followed a decade later.

In the German university town of Iena in 1884 three men set about revolutionising the world of glass. The story begins with Carl Zeiss, an engineer who had moved to Iena 38 years earlier. At first he earned a living repairing instruments for the university laboratories, then he opened a small workshop making optical equipment, principally microscopes. Yet Zeiss himself was not a scientist, so in 1866 he joined forces with Ernst Abbe, a trained physicist and researcher in the field of optics, with the aim of developing new products. Gradually the team began to win a reputation. With Zeiss's encouragement, Abbe designed increasingly powerful microscopes, but eventually ran up against the limitations of the glass available commercially at the time.

At that point Otto Schott entered the picture. In 1881, having obtained a doctorate in the chemistry of glass-making, he sent a sample of his glass, based on lithium, to Abbe, who was deeply impressed by the homogeneity of the product. Learning that Schott was working on his own from his parents' home in Aachen, Abbe invited him to join the team in Iena. From 1882, Schott began systematically to explore the results that could be achieved with different metals and acids, producing a number of optical glasses to add to the repertoire the firm already offered. In his first three years in Iena, Schott developed more than 20 new varieties of glass, mostly destined for use in optical instruments.

Ultra-resistant glass

In 1884 Schott was experimenting with boric acid when he hit upon borosilicate glass. Made up of boron, silicon and oxygen, his chance invention proved resistant to temperatures as high as 400°C (750°F), as well as to sudden shocks and contact with chemical products. It would prove useful not just in chemical laboratories but also in the manufacture of light bulbs, which Thomas Edison had recently patented in the USA. Soon the world was beating a path to the Iena team's door. In those innovative Industrial Revolution days, it was a short step from laboratory to mass production. The various stages in glass production were mechanised, and the traditional crucible furnaces were replaced by continuous-fusion gas ovens of the type that the Siemens firm had introduced in Germany in 1867. The Zeiss factories were soon working flat out. Carl Zeiss himself died in 1888, in the early days of their triumph. Three years later Abbe created the Carl Zeiss Foundation, giving it the

1851, the Paris World's Fair of 1889 provided a showcase for glass and its new importance as a construction material, notably in the *Galerie des Machines* on the Champ-de-Mars, which was built entirely of metal and glass. Yet the material still remained fragile. That changed from 1893 on, when a French engineer named Léon Appert had the idea of trying out on glass a concept that had been successfully used to reinforce concrete. He inserted into the molten glass a thin grid of metal wires 0.6–0.9mm in diameter, in a mesh of between 10mm and 30mm spacing. The presence of the wire grill made little difference to the quality of transparency, but glass reinforced in this way was four times stronger than normal glass; in addition, when a sheet did break, it did so into tiny morsels without sharp edges.

responsibility for managing a new, highly profitable firm named Jaener Glaswerk Schott & Genossen, today known as Schott AG.

The birth of Pyrex

In time the glassware developed at Iena would find some unexpected applications. In 1913 the wife of an employee of the American Corning firm, which made borosilicate glass, had the idea of trying out the bowl-shaped lower section of a heat-resistant lamp as a cooking vessel. She put a cake in the bowl, and it came out perfectly baked. Glass had not previously been used to make cooking utensils, but Corning set to work on the idea and two years later launched an entire range of ovenproof ware under the Pyrex brand name.

The work at Iena also paved the way for improvements in optical glass. The lenses used for headlamps, traffic lights and railway signals all benefited from Abbe and Schott's work.

Reinforced glass

Following a trail blazed by the Crystal Palace at London's Great Exhibition of

Glass factory
Founded in Paris in 1875, the Appert glassworks (right) were managed in later decades by Léon Appert, who invented reinforced glass in 1893.

GALLÉ AND THE SCHOOL OF NANCY

After studying glass-making in Germany, Émile Gallé began creating vases and lamps in new and original shapes, inspired by nature and decorated with plant motifs like the chrysanthemums on this lamp (left). To achieve his effects he adopted recent innovations in the industry, including the use of hydrofluoric acid for etching and the addition of novel colourants. Recognised as a master of the Art Nouveau style, he came together with other leading practitioners, including Louis Majorelle and the Daum brothers, to make Nancy in the Lorraine region of eastern France a centre of the movement. Gallé's American contemporary was Louis Comfort Tiffany, whose iridescent glassware and stained-glass windows became celebrated around the world as masterpieces of the decorative arts.

THE THERMOS SAGA

In 1892 when the Scottish physicist James Dewar invented a receptacle capable of maintaining liquids at a constant temperature, he never dreamed that he would miss out on a fortune that could have helped finance his life's work on very low temperatures. His flask consisted of two separate bottles, one set inside the other, separated by a vacuum that acted as an insulator. A coating of silver on the sides of the glass reduced the transfer of heat through radiation. Dewar's flask initially attracted the attention of other scientists, but in 1904 a German glass-blower noticed that milk intended for his baby remained warm when stored in this obscure research tool. With the help of a colleague, he set about commercialising the flask under the Thermos tradename, Dewar himself having neglected to patent his discovery. From 1907 on the Thermos flask gained worldwide publicity when it was used by Peary and Shackleton on their expeditions to the polar regions. In the years leading up to the First World War, the flasks accompanied the Wright brothers in their aircraft and Count von Zeppelin in his dirigible. Marketed in bottle shape in the 1920s and 30s, the Thermos preceded refrigerators as a receptacle for conserving foodstuffs. From 1966 steel replaced glass in the flasks' manufacture, making them unbreakable. Meanwhile Dewar's flask continued to find a use in laboratories, where its efficiency was increased by the incorporation of total vacuums. Its principal used in science today is for storing liquid nitrogen at temperatures down to -196°C – low enough to preserve tissues, cells and embryos for scientific research.

Inside a flask *This cutaway drawing shows the two separate receptacles of a thermos vacuum flask, one inside the other.*

Bulletproof 'glass'
The material used to deflect gunshots is in fact mostly laminated polycarbonate.

Shatterproof glass was soon put to use in floors and roofing. In the first years of the 20th century it could also be found in the windshields of trains, trams and automobiles.

By inventing reinforced glass Appert had introduced the concept of lamination – the bonding together of thin sheets in a layered structure – that underlies all so-called 'intelligent' modern glasses. Triple-layer laminated glass was patented by the French chemist Edouard Benedictus in 1909 and became available commercially from 1920 on.

Today, up to 13 layers may be incorporated in a single sheet of glass, each one only a few micrometres thick. Metals, metallic oxides, plastics and polycarbonates can all be used to darken glass as required, permitting only part of the visible spectrum to pass through; alternatively, glass surfaces can serve to display an image. Modern reinforced glass also has improved mechanical qualities, making it resistant even to severe impacts. It is used today to produce insulating materials, while optical fibres have in recent years helped to transform telecommunications.

The electric toothbrush 1884

Scott's 'electric' toothbrush
Under a rather broad interpretation of 'electric' the doctor also advertised electric hairbrushes, an electric corset and an electric curry comb for horses.

The first supposedly electric toothbrush was patented in 1884 by an Englishman, Dr George A Scott. It was one of a range of brushes that he advertised and was not, in fact, powered by electricity at all. What it did have was a magnetised iron rod in the handle that purportedly generated an 'electro-magnetic current' beneficial for dental health. In fact it was the 1950s before a genuinely electric device was invented. Even then, sales only took off after General Electric introduced a cordless, rechargeable version in 1961. Customers had apparently been turned off previous models by the fact that they were connected to the mains.

Coca-Cola 1885

In 1885 a chemist from Atlanta, Georgia, named John Styth Pemberton concocted a new drink made from, among other ingredients, alcohol, kola nuts, sugar, caffeine and coca leaves, the raw source of cocaine. He sold it as a tonic under the name of Pemberton's French Wine Coca. Soon afterwards the state of Georgia introduced temperance legislation, and Pemberton responded by removing the alcohol. In 1887 his accountant suggested changing the name to Coca-Cola and devised the familiar logo. Pemberton subsequently sold the rights to a businessman named Asa Griggs Chandler.

Until 1916 Coca-Cola was mostly sold as a health drink at soda fountains for 5 cents a glass, although it had been available in bottled form from 1894. The familiar, shaped bottle with a bulge around the middle dates from 1915, when the firm launched a competition for a container that 'a person could recognise even if they felt it in the dark'. The winner was the Root Glass Company, whose designer, Earl R Dean, took his inspiration from an encyclopedia illustration of a coca pod. The exact recipe for the drink remains a closely guarded secret to this day. The active coca extract was removed in 1904.

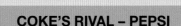

Iconic bottle
The designer was reportedly rewarded with the offer of a job for life with the bottling company. His design was familiar by the time this delivery truck was seen in New Orleans in 1929.

COKE'S RIVAL – PEPSI

In 1893 a North Carolina pharmacist named Caleb Bradham invented a drink based on caramel, sugar, aromatic extracts and carbonated water intended to soothe stomach pains and provide energy. In 1898 he settled on the name Pepsi-Cola, trademarked in 1903. In 1931 Charles G Guth, who had by then bought up the business, sought to take trade away from its rival Coca-Cola by temporarily selling double-sized bottles for the same price as standard bottles of Coke. The strategy paid off, and Pepsi continues to rival Coke to this day.

An electrifying genius

Nikola Tesla is generally regarded as one of the great geniuses of his day. He was responsible for more than 700 inventions in the field of electricity and he introduced the principle of alternating current as the basis of the electrification of modern societies.

Portrait of genius
This photograph of Nikola Tesla was taken in a studio in 5th Avenue, New York, in about 1915, when the remarkably young-looking physicist was approaching 60 years old.

'Were we to seize and eliminate from our industrial world the results of Mr Tesla's work, the wheels of industry would cease to turn, our electric cars and trains would stop, our towns would be dark, our mills would be dead and idle.' The claim, made by a former president of the American Institute of Electrical Engineers, is no exaggeration: today Nikola Tesla is universally recognised as a visionary genius, the principal champion of alternating current (AC) in the 'war of currents' that divided the industrialised world at the end of the 19th century. His achievements in this field dwarf even the 700-plus inventions chalked up to his name, for between 1882 and 1893 he succeeded in proving the superiority of the AC system over the direct current (DC) alternative as a means of electrifying the world.

Destined for the clergy

Born to Serbian parents in Smiljan, Croatia, in 1856, Tesla became a naturalised American citizen in 1891. His father was a pastor in the Serbian Orthodox Church and had intended his son for an ecclesiastic career, but he eventually allowed Nikola to follow his own inclinations, won over by his astonishing intellectual gifts. He was also no doubt aware of the signs of psychological instability that Tesla already displayed in the course of a sickly childhood.

While his views were humanistic and pacifist, Tesla avoided human company in the most literal sense, fearing microbial contagion as well as psychological intrusion. He took refuge in the powerful and hermetic world of his own imagination. People who were close to him described him as living in a perpetual trance, with tremendous powers of visualisation that made the inventions he imagined seem already real to him, issuing from his mind fully formed.

Tesla enrolled in a polytechnic college in the Austrian city of Graz at the age of 17 to study engineering. There he found his true vocation in life. Confronted with a Gramme machine – a type of direct-current dynamo – he was as sceptical as he was impressed by its performance, considering that it gave off too many sparks to be truly efficient. He was

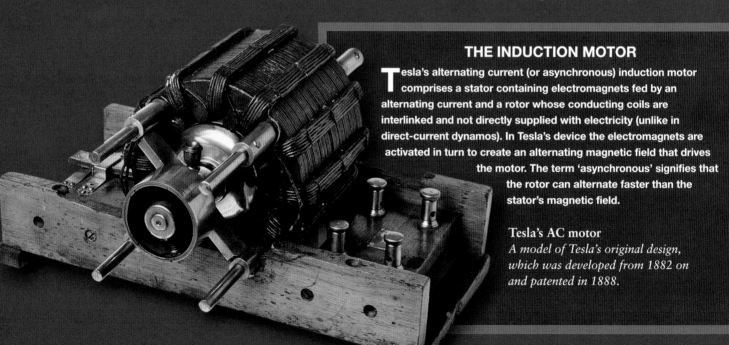

THE INDUCTION MOTOR

Tesla's alternating current (or asynchronous) induction motor comprises a stator containing electromagnets fed by an alternating current and a rotor whose conducting coils are interlinked and not directly supplied with electricity (unlike in direct-current dynamos). In Tesla's device the electromagnets are activated in turn to create an alternating magnetic field that drives the motor. The term 'asynchronous' signifies that the rotor can alternate faster than the stator's magnetic field.

Tesla's AC motor
A model of Tesla's original design, which was developed from 1882 on and patented in 1888.

A coil for CERN

This massive coil (right), constructed in 1970 for the European Centre for Nuclear Research, was built on the principle of the Tesla motor. The particle detector is equipped with a superconducting magnet that creates a significant magnetic field, enabling scientists to study the trajectory of particles and so to analyze their mass and electrical charge. The small coil above is from Tesla's alternator.

convinced he could do better and dreamed of creating a simpler, more effective device. In effect, the alternating-current induction motor would be his greatest invention.

Early triumphs and setbacks

The idea for this profoundly innovative machine, which marked the start of the AC revolution, first came to him in February 1882, while he was working for the National Telephone Company in Budapest. He moved to Paris to join the French subsidiary of Thomas Edison's Continental Edison Company, then in 1883 he was transferred to Strasbourg, where he demonstrated a prototype of the motor. Although he did not receive any financial reward for his invention, the director of the Strasbourg branch realised his potential and in 1884 dispatched him to join Edison in the USA. Tesla arrived full of enthusiasm but was quickly disillusioned; finding Edison more interested in profit than in the scientific worth of his inventions, he soon ended up in conflict with his employer.

The war of currents

Tesla resigned from Continental Edison in 1885, opening his own Tesla Electric Light & Manufacturing Company the following year. The 'war of currents' now broke out into the

UNDERPAID AND UNDERVALUED

Tesla had good reason for falling out with his employer Thomas Edison, who not only patented many of the younger man's inventions in his own name but also short-changed him financially. By his own account, Tesla understood that Edison had promised him a reward of $50,000 for redesigning his direct-current generators, but when he asked for the money he was supposedly told, 'Tesla, you don't understand our American humour'. To make matters worse, he was refused a pay rise from $18 to $25 a week.

open. Tesla flung himself into the production of AC instruments – generators, motors, induction coils, transformers, condensers. He also produced arc lights and incandescent bulbs and took out more than 40 patents. The year 1888 saw the publication of a ground-breaking article entitled *A New System of Alternating Current Motors and Transformers*. By that time he had sold his patents to the industrialist George Westinghouse, who became his collaborator in propelling first the United States and then the whole world into the AC era, thereby effectively bringing Edison and his DC aspirations down to earth.

WHY AC?

The main advantage of alternating current (AC) over direct current (DC) is that it can easily be transformed to higher or lower voltages. Power stations generate high voltages, since these are more efficient for transmitting electricity over long distances. The high voltages are then easily reduced to safer low voltages for use in the home.

Wardenclyffe Tower
Located on Long Island, New York, the tower (below) was built for Tesla between 1901 and 1903. He hoped to use it to prove that it was possible to transmit electrical energy across the Atlantic without the aid of wires. The project was never realised.

Mountain retreat
Tesla (above) photographed at his experimental station outside Colorado Springs in the Rocky Mountains. He was working on ways to transmit power over long distances and also seeking to communicate with other planets.

An indefatigible inventor

The incredible machines that the Tesla–Westinghouse team developed came to international attention at the Chicago World Fair of 1893. Two years later Tesla designed the first large-scale hydroelectric power station, built on the alternating-current principle, to tap the massive power resources of the Niagara Falls. By that time, he and Westinghouse had in effect won the war of currents. Yet Tesla never let the conflict with Edison absorb all his energies. In 1887, for example, he was also working on cathode-ray tubes and came close to beating Wilhelm Röntgen to the discovery of X-rays.

In 1891 Tesla demonstrated that electrical energy could be transmitted from a generator to a light bulb without passing through a wire. This was a consequence of the properties of electromagnetic waves, which had first been brought to light by Hertz in 1884. In 1899 Tesla managed to illuminate 200 incandescent bulbs of 50W each over a distance of 26 miles in the mountains near Colorado Springs.

A TORMENTED SPIRIT

Tesla's genius also found expression in non-scientific fields, notably in languages. In addition to English he spoke Serbo-Croatian, Czech, French, German, Hungarian, Italian and even Latin. Yet his restless spirit was troubled by psychological problems. In his autobiography he described his obsessive-compulsive behaviour: 'All repeated acts or operations I performed had to be divisible by three and if I missed I felt impelled to do it all over again, even if it took hours.'

In 1895 Tesla took out a patent on a remotely controlled torpedo boat. Two years later he conceived the basic principles of the radio and, three years after that, of radar. He dreamed of establishing a worldwide communications network, but the idea was too revolutionary to be realised at the time.

Tesla remained an innovator right up until his death on 7 January, 1943, at the age of 86. Shortly before he died he had been working for the US Air Force on the development of a 'death ray' – apparently some sort of particle-beam gun capable of destroying entire cities that he hoped would render war obsolete. Yet for all his creative energy and achievements, Tesla was never a public figure in the way that his rival, Edison, was. He despised financial rewards and honours, reportedly turning down the Nobel prize for physics in 1915 because he would have had to share it with Edison. It was only at his publisher's insistence that he was persuaded to produce an autobiographical account of his work, *My Inventions*, which was first published in 1919. Among many posthumous honours for his work, Tesla's name was adopted for the international SI unit used to measure magnetic flux density, a fitting recognition for the man who did more than almost any other to prepare the way for 20th-century technology.

ALTERING THE CLIMATE

In 1899 Tesla made what he described as his greatest discovery: terrestrial stationary waves. The idea was that the Earth itself could serve as an enormous electrical conductor. In the journal of his inventions he proposed using electromagnetic waves to study and act upon the ionosphere, the uppermost part of the atmosphere located 50 to 1000km (30–600 miles) above the planet's surface, which contains the electrically charged particles known as ions. Astrophysicists worked on the idea throughout the 20th century, and it also proved of interest to the US military, whose controversial High Frequency Active Auroral Research Programme might, according to some critics, be used to disrupt the climate of whole continents.

The death ray
Tesla's proposed particle-beam weapon inspired the creators of the Star Wars *films (top right) and also attracted the attention of the US military in the context of the Strategic Defence Initiative (right) launched by President Ronald Reagan in 1983.*

The bicycle revs up

With its wooden frame, a seat like a horse's saddle and stabilising wheels at its sides, the prototype of the modern-day motorbike looked deceptively like a child's toy. Even then, though, its small, sputtering engine could propel a rider along at a heady 12mph.

Early motorcycle
A new model being tested in about 1910 (right) at the Daimler factory in Stuttgart, where the first motorcycle had been produced 25 years earlier.

In 1885, when he constructed the first motorcycle for the industrialist Gottlieb Daimler, the German engineer Wilhelm Maybach had little idea that he was about to revolutionise travel on two wheels. The device was intended merely to test the performance of a four-stroke petrol motor destined for a wild project Daimler was working on – to build a horseless wagon capable of carrying passengers. These 'automobiles', as they would be known, became a reality a decade later.

Motorising the velocipede

By that time pedal-powered bicycles, which had evolved from the velocipede invented earlier in the 19th century, were a common sight. The heavy metal frames demanded vigorous pedalling by the rider, but they were cheap and easy to come by and proved to be ideal experimental platforms for a handful of inventors eager to make cycling less like hard work. One such was the Frenchman Louis Guillaume Perreaux, who in 1868 took out a patent on a high-speed velocipede to which he had attached a steam engine. Other attempts to apply steam power to bicycles were made in the USA and elsewhere in Europe, but all came to nothing. When it came to two-wheeled vehicles, the steam engine had a fatal flaw: it was simply too big and too cumbersome for the amount of space available.

In marked contrast, combustion engines were self-contained and compact, with an excellent weight-to-power ratio. These advantages came to the fore in the final decade of the 19th century as literally hundreds of bicycle manufacturers launched themselves into the production of motorised vehicles, with varying degrees of ambition and success. In 1885 the German engineer Siegfried Bettman arrived in Coventry from Nuremberg to produce motorcycles and in 1886 bought the name Triumph, which launched its first machines in 1889 and went from strength to strength. During World War One more than 30,000 Triumph machines were supplied to the Allies.

The thrill of speed

The very early motorcycles were mostly bought by wealthy individuals who could afford to enjoy the thrill of adventure, even if it involved a noisy and uncomfortable machine that frequently broke down. These pioneers,

---

A FAILED EXPERIMENT IN MASS PRODUCTION

In 1894 the German firm Hildebrand & Wolfmüller made the first attempt to manufacture production-line motorcycles. The *Motorrad* had an impressively large 1490cc engine and a top speed of 28mph (45km/h). A few hundred examples were built, but the engine had a number of design flaws, made worse by the absence of a clutch, and proved vulnerable to dust and damp, leading to frequent breakdowns. Complaints poured in and the firm was unable to meet all the demands for reimbursements. Bankruptcy followed in 1897.

Sunday outing
When this photograph was taken in Germany in 1899 (above), motorcycling on two or three wheels was a leisure activity for the rich and adventurous.

for whom motorcycling was purely a leisure activity, saw their toys become much more sophisticated once the 20th century dawned. What had been bicycles with motors added almost as an afterthought were now transformed into automobiles on two wheels with engines whose increasing horsepower finally did away with the need for supplementary pedals.

The engine, which had previously sometimes been placed either on the front or rear mudguard, now found a more or less definitive home in the centre of the frame. Gearboxes and clutches became common, increasing the vehicles' traction. Transmission belts put in an appearance, and the old driving wheel was replaced by an altogether more reliable metal chain that did not slip or stretch.

Maximum speeds began to increase rapidly. By 1912 the Norton Old Miracle, made in Wolverhampton, managed 80mph (130km/h). Motorcycle riding became something of a feat of acrobatics on roads that were often unsurfaced, and the springs under the saddle rarely sufficed to absorb the jolts any better than the rubber tyres, used on bicycles since 1888. Consequently, the first suspension systems were introduced at this time. By the

De Dion-Bouton tricycle
Made in Coventry for the British market, this 1898 tricycle model would shortly be replaced by a two-wheeled version.

THE FIRST MOTORCYCLE RACES

The first race events were haphazard affairs in which two, three and four-wheeled vehicles sometimes competed against one another. In May 1895 three of five competitors in an Italian race from Turin to Asti and back rode Hildebrand & Wolfmüller motorcycles. The following year the Paris–Mantes race in France saw six tricycles lined up against one motorcycle. The first recorded event exclusively for two-wheelers took place at Richmond, Surrey, in 1897.

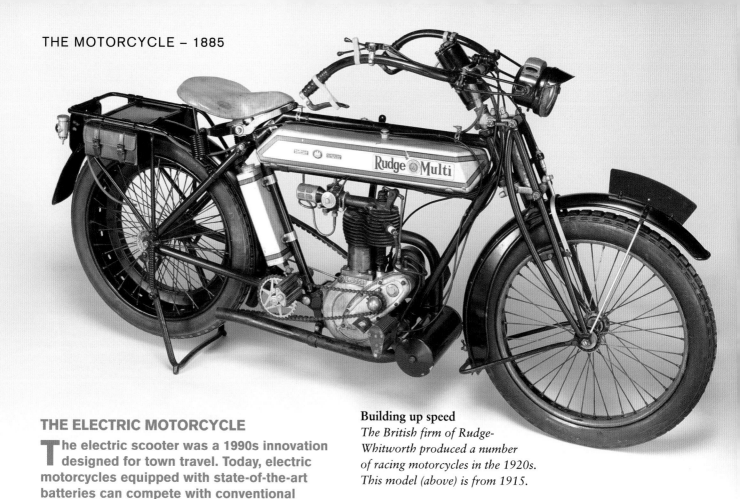

THE ELECTRIC MOTORCYCLE

The electric scooter was a 1990s innovation designed for town travel. Today, electric motorcycles equipped with state-of-the-art batteries can compete with conventional 250hp models. Experimental prototypes have already reached 150mph (240km/h).

Building up speed
The British firm of Rudge-Whitworth produced a number of racing motorcycles in the 1920s. This model (above) is from 1915.

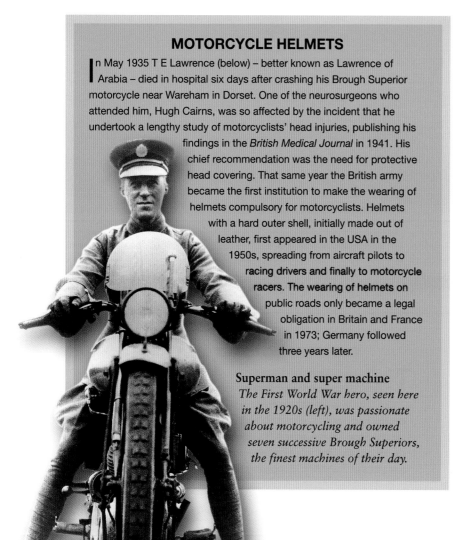

MOTORCYCLE HELMETS

In May 1935 T E Lawrence (below) – better known as Lawrence of Arabia – died in hospital six days after crashing his Brough Superior motorcycle near Wareham in Dorset. One of the neurosurgeons who attended him, Hugh Cairns, was so affected by the incident that he undertook a lengthy study of motorcyclists' head injuries, publishing his findings in the *British Medical Journal* in 1941. His chief recommendation was the need for protective head covering. That same year the British army became the first institution to make the wearing of helmets compulsory for motorcyclists. Helmets with a hard outer shell, initially made out of leather, first appeared in the USA in the 1950s, spreading from aircraft pilots to racing drivers and finally to motorcycle racers. The wearing of helmets on public roads only became a legal obligation in Britain and France in 1973; Germany followed three years later.

Superman and super machine
The First World War hero, seen here in the 1920s (left), was passionate about motorcycling and owned seven successive Brough Superiors, the finest machines of their day.

eve of the First World War motorcycles had largely taken on their modern form, along with most of the features familiar today.

Putting motorbikes to work

In the 1920s cars were still too expensive for most people. Motorcycles offered a cheaper alternative that could travel as far and as fast, with some machines already capable of touching 125mph (200km/h). No longer just recreational vehicles for the rich, they became an essential means of transport for whole swathes of the middle class and better-off working class. Often a sidecar was attached to allow passengers to travel in greater comfort. The motorcycling phenomenon became even more apparent after the Second World War when mass production, particularly of less powerful machines, made travel on two wheels genuinely cheap. For a time it was the most economical form of motor transport.

Then, in the 1960s, motorcycles ran up against fierce competition from both ends of the spectrum. On the one hand mopeds, with engine capacities of 90cc or less, were even cheaper and well adapted to town traffic; on the other hand cars had become less expensive as production volumes grew, putting them within the economic reach of far more people. Two-wheeled travel lost out in comparison

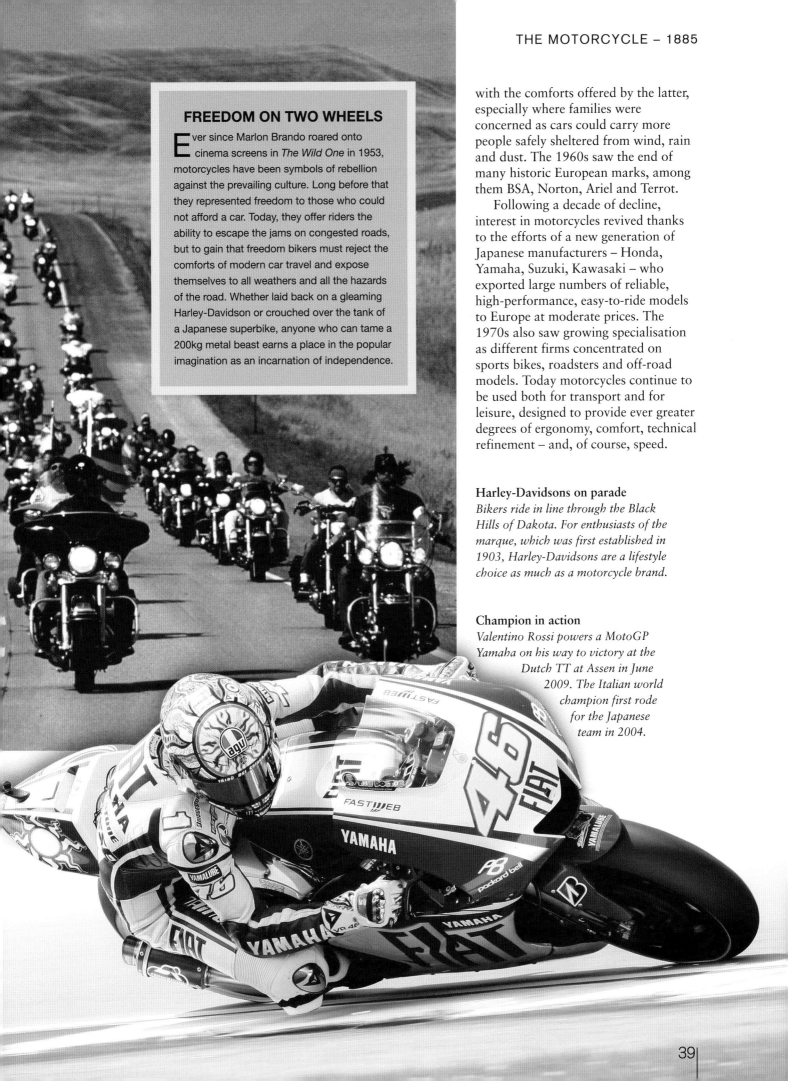

FREEDOM ON TWO WHEELS

Ever since Marlon Brando roared onto cinema screens in *The Wild One* in 1953, motorcycles have been symbols of rebellion against the prevailing culture. Long before that they represented freedom to those who could not afford a car. Today, they offer riders the ability to escape the jams on congested roads, but to gain that freedom bikers must reject the comforts of modern car travel and expose themselves to all weathers and all the hazards of the road. Whether laid back on a gleaming Harley-Davidson or crouched over the tank of a Japanese superbike, anyone who can tame a 200kg metal beast earns a place in the popular imagination as an incarnation of independence.

with the comforts offered by the latter, especially where families were concerned as cars could carry more people safely sheltered from wind, rain and dust. The 1960s saw the end of many historic European marks, among them BSA, Norton, Ariel and Terrot.

Following a decade of decline, interest in motorcycles revived thanks to the efforts of a new generation of Japanese manufacturers – Honda, Yamaha, Suzuki, Kawasaki – who exported large numbers of reliable, high-performance, easy-to-ride models to Europe at moderate prices. The 1970s also saw growing specialisation as different firms concentrated on sports bikes, roadsters and off-road models. Today motorcycles continue to be used both for transport and for leisure, designed to provide ever greater degrees of ergonomy, comfort, technical refinement – and, of course, speed.

Harley-Davidsons on parade
Bikers ride in line through the Black Hills of Dakota. For enthusiasts of the marque, which was first established in 1903, Harley-Davidsons are a lifestyle choice as much as a motorcycle brand.

Champion in action
Valentino Rossi powers a MotoGP Yamaha on his way to victory at the Dutch TT at Assen in June 2009. The Italian world champion first rode for the Japanese team in 2004.

Hot-metal typesetting speeds up printing

Newspapers became democratic in the course of the 19th century as new modes of transport improved distribution and the rotary press, in use from 1847, increased print runs making them cheaper. The next advance was to accelerate the typesetting process.

A rival system
Linotype was the preferred choice of the daily press, but Monotype gained a place in quality printing, mostly of books. The Monotype system employed two separate machines (below and below right).

Before 1886 typesetters still assembled by hand the individual characters making up each block of text to be printed, much as they had done since printing was invented. They took the letters, one by one, from the cases in which they were kept then patiently placed, or composed, them in line. The process was time consuming and costly – so much so that the search for an alternative had been going on for some time, with systems being tested as early as 1822. But it was not until 1886 that the process was transformed with the arrival of Linotype, invented by Ottmar Mergenthaler, a German-born American clockmaker. His hot-metal typesetting machine, the first of its kind, removed the need to handle individual characters. Instead, it directly moulded entire lines of type, setting copy at a rate of 6,000 characters an hour, rather than the 1,000 to 1,300 characters achieved manually.

Goodbye to movable type

Mergenthaler's invention was an imposing machine more than 2m tall. At a time when typewriters themselves were still new, it featured a 90-character keyboard on which the characters were arranged by frequency

Line of setters
Linotype operators at work in the composing room of the New York Herald *in 1902.*

of use. Each stroke made by the operator selected a matrix – a metallic letter mould – from one of a number of type magazines, each of which held a particular font; the machine then assembled the chosen characters in lines of text. Thanks to an ingenious distribution system, the matrices were mechanically returned to their original places. The machine automatically justified the text, inserting spaces between words to ensure that each line was the same length. Individual lines or 'slugs' of type

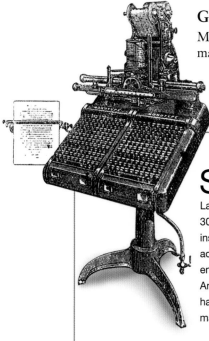

MONOTYPE

Shortly after the introduction of Linotype, a rival typesetting system became available in the shape of Monotype, patented by the American Tolbert Lanston in 1887. It involved two separate machines: a keyboard with about 300 keys (left), which produced perforations in a paper tape that then provided instructions to operate the typesetting device (right). Lanston's invention had several advantages over Linotype. It employed a superior alloy to mould individual characters, enabling it to produce higher-quality work. Corrections could be made more easily. And separating the keyboard from the typesetter helped to protect workers from the harmful vapours released by the molten lead used to shape characters within the matrices. In time Linotype machines too would be operated by perforated strips.

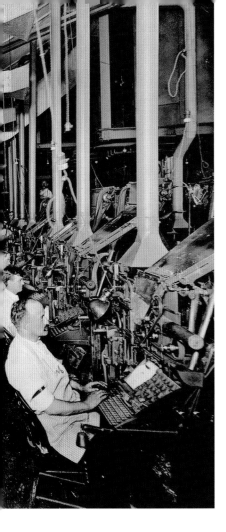

LINE O'TYPE

Legend has it that on the very day in 1886 when Mergenthaler demonstrated his new machine at the offices of the *New York Tribune*, Whitelaw Reid, the impressed editor-in-chief, exclaimed: 'Ottmar, you've done it again! A line o'type!' Mergenthaler subsequently adopted the phrase as the name for the new machine.

A POETIC PRECURSOR

In 1845 the French poet Gérard de Nerval patented a typesetting system that involved composing lines of text in a single block.

were then assembled in a forme that matched the dimensions of the finished page, ready for printing.

The birth of a new profession

Linotype was first used at the *New York Tribune* in 1886. In combination with the increased mechanisation of printing, it dramatically speeded up the production of newspapers. Big-circulation dailies made their appearance and the newspaper industry became a highly competitive mass-market business. A British subsidiary of Mergenthaler's firm was set up in Manchester in 1895, proving successful enough to open up a second factory in Altrincham four years later.

Attracting a number of imitations over the years, Mergenthaler's machine created a whole new profession: linotypists would be familiar figures in print works for almost a century to come. The machines themselves remained in use until the 1970s, when they were eventually made redundant by the advent of 'cold-type' photosetting.

THE BIRTH OF THE POPULAR PRESS

Media historians trace the birth of the popular press in Britain to the foundation of the *Daily Mail* by Alfred Harmsworth (later Lord Northcliffe) in 1896. One of 14 children of a cash-strapped barrister, Harmsworth made his fortune with a series of popular magazines before launching into the newspaper market in 1894 with the purchase of the *London Evening News*. Helped by snappy news reporting, the introduction of popular features such as competitions and a selling price that was half that of its rivals, the new paper soon proved a commercial success. It offered a populist formula that contrasted with the text-heavy serious newspapers common across Europe at the time (above).

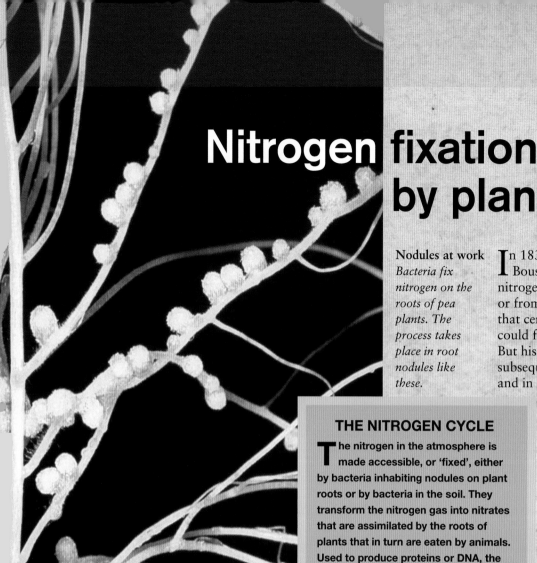

Nitrogen fixation by plants 1886

Nodules at work
Bacteria fix nitrogen on the roots of pea plants. The process takes place in root nodules like these.

In 1837 the French chemist Jean Baptiste Boussingault set out to discover whether the nitrogen present in plants came from the soil or from the air. His experiments convinced him that certain plants, such as beans and lupins, could fix the gas from atmospheric sources. But his one-time assistant Georges Ville subsequently reached the opposite conclusion, and in 1855 the French Academy of Sciences came down on Ville's side.

In fact, both men were right, but this only became apparent after 1886 when Mikhail Voronin and Hermann Hellriegel identified bacteria capable of assimilating nitrogen in nodules on the roots of leguminous plants. In 1891 Sergei Vinogradski established that plants fix nitrogen both from the air and from the soil. Other scientists would go on to elucidate the mechanisms involved and to develop the new science of soil fertilisation, the seeds of which had first been sown by Justus von Liebig in the 1840s.

THE NITROGEN CYCLE

The nitrogen in the atmosphere is made accessible, or 'fixed', either by bacteria inhabiting nodules on plant roots or by bacteria in the soil. They transform the nitrogen gas into nitrates that are assimilated by the roots of plants that in turn are eaten by animals. Used to produce proteins or DNA, the nitrogen eventually returns to the earth or the atmosphere as nitrates through the decomposition of organic matter.

Germanium 1886

The Russian biologist Dmitri Mendeleev introduced the periodic table of elements in 1869, but the space he allotted for element No.32 remained empty until 1886. That year, the German chemist Clemens Winkler was examining agyrodite, a newly discovered mineral containing silver and sulphur, when he identified a metalloid element that closely matched the properties predicted for the missing occupant of slot 32. The new element was named germanium. There were no known uses for it at the time, but it played a part in the creation of the first transistor in 1947, a role later usurped by silicon. Today about 100 tonnes of germanium are produced annually for use in the fabrication of fibre optics and in a few other specialised applications, including the production of some electronic components and night-vision lenses.

X-ray view
The first application of germanium was in semiconductors like this transistor.

Esperanto 1887

In 1887 Ludwik Zamenhof published a book introducing an international language he had devised under the pseudonym Doktoro Esperanto, 'Doctor Hopeful'. Zamenhof was 28 years old at the time, a Russian Jew from the small town of Bialystok on the edge of the tsarist empire in what is now Poland. From an early age he was convinced that the diversity of languages was the main 'source of dissension in the bosom of the human family', and so he worked out a simple new language to facilitate better communication between different peoples. Each letter in every word was always pronounced in the same way; grammar was summarised in 16 invariable rules. Esperanto was recognised as an independent language by UNESCO in 1954 and by 1970 it was on the curriculum of 15 universities worldwide, rising to 115 in 1987. No-one knows how many people speak it today, but in 1987 a centenary congress brought together 6,000 participants from 72 different countries.

Promoting Esperanto *The first congress for speakers of Esperanto was held in the French port of Boulogne-sur-Mer in 1905. This photograph shows a promotion from 1910.*

The duplicating machine 1888

In 1888 a Hungarian named David Gestetner first demonstrated a machine capable of producing six copies of a document a minute. His Cyclograph transferred text through the use of stencils: ink passed through perforations in waxed paper to print copies on the blank sheets placed underneath. Seven years earlier Gestetner had developed an improved method of cutting handwritten stencils using a hand-held device he called the Cyclostyle pen. Now he had adapted that technology for the dawning typewriter era, introducing a flatbed printing surface and a toothed wheel. At roughly the same time, Thomas Edison was working on a rival system, the mimeograph machine. Rotary duplicators appeared in the first decade of the 20th century under the Gestetner and Roneo trademarks; they were supplanted in their turn by the photocopier from the 1960s.

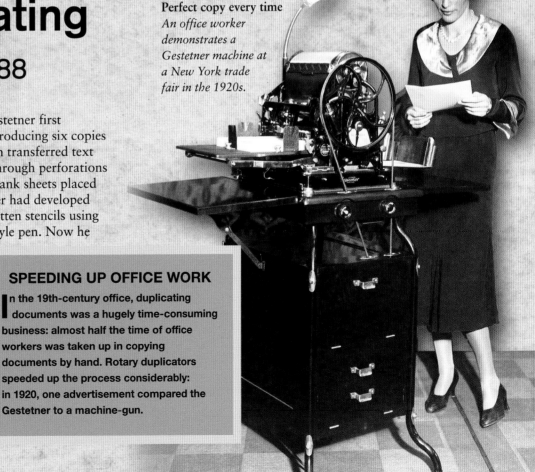

Perfect copy every time *An office worker demonstrates a Gestetner machine at a New York trade fair in the 1920s.*

SPEEDING UP OFFICE WORK

In the 19th-century office, duplicating documents was a hugely time-consuming business: almost half the time of office workers was taken up in copying documents by hand. Rotary duplicators speeded up the process considerably: in 1920, one advertisement compared the Gestetner to a machine-gun.

Analysing how the body moves

E tienne-Jules Marey deserves an honoured place in the pantheon of scientific visionaries. A maverick scholar, he undertook important work in several different fields, evoking something of the inventiveness of a modern-day Leonardo da Vinci.

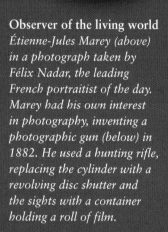

Observer of the living world
Étienne-Jules Marey (above) in a photograph taken by Félix Nadar, the leading French portraitist of the day. Marey had his own interest in photography, inventing a photographic gun (below) in 1882. He used a hunting rifle, replacing the cylinder with a revolving disc shutter and the sights with a container holding a roll of film.

In the 1860s, having completed a doctoral thesis on the circulation of the blood, Étienne-Jules Marey decided to devote himself to further research on physiology. His first invention was a portable sphygmograph, a device for recording blood pressure on paper.

As a rising star of the discipline, in 1869 Marey was appointed Professor of the Natural History of Organised Bodies at the Collège de France in Paris; he was elected to the French Academy of Sciences nine years later. In 1880 he created a physiological laboratory in the grounds of the Parc des Princes velodrome on the outskirts of Paris, where he devised and made equipment and undertook experiments. He was particularly interested in motion, of animals as well as humans. The guiding idea that steered his work was that movement needed to be observed and recorded in detail if it was to be studied, analysed and measured.

Inventing chronophotography

In 1874 the astronomer Jules Janssen had used a photographic gun to record the transit of Venus across the Sun. Taking this as his inspiration, Marey determined to develop his own version to photograph movement. In 1882, while in Naples where he spent part of each year, he produced a device that took 12 exposures a second and used it to capture the flight of birds. That spring he demonstrated his invention to the Academy of Sciences in Paris, revealing successive images of a seagull on the wing, but the pictures from this first attempt were small and blurred.

The same year Marey developed the fixed-plate chronophotographic camera, inspired this time by the work of Britain's Eadweard Muybridge. This new camera featured a slotted disc revolving in front of the photographic plate, which made it possible to capture many successive images – it was, in effect, the first rotary disc shutter. At first Marey used his invention to take stop-motion shots of moving figures dressed in white against a black background. Later he substituted dark outfits that made his subjects all but invisible, so that only their silhouettes and specific highlights deliberately placed at joints or along the limbs registered on the plates. Using this system Marey and his colleague Georges Demenÿ spent months 'chronophotographing' the movements of athletes and animals.

THEORIST OF AERONAUTICS

H aving abandoned his early belief that the secret of human flight lay in the invention of machines provided with beating wings, Marey became an avid promoter of the aeroplane. In 1899 he built a prototype wind tunnel to study how air flowed around objects and obstacles of different shapes.

Toward cinema

In time Marey replaced the fixed plate with a moving paper strip, soon abandoned in favour of the rolls of celluloid film that George Eastman had recently invented for Kodak. In doing so he moved chronophotography close to the dawning world of cinematography, which would ultimately replace it. Marey in fact made contact with the Lumière brothers, whose contribution to the birth of cinema would come to overshadow his own. In the event, Marey disdained the idea of his invention being exploited simply for recreation and refused to allow his work to be shown in animated form except for purposes of scientific analysis.

Gripped by a wish to 'see the invisible', in 1891 Marey adapted his photo gun to operate through the eyepiece of a microscope. In this way he managed to obtain several series of images of various marine micro-organisms, including zoospores and algal cells, as well as of the movement of blood in the capillaries.

Marey liked to describe himself as an 'engineer of the living world' and in 1894 he summed up his researches as a scientist in a work entitled simply *Le Mouvement* ('Movement'). He died in Paris in 1904.

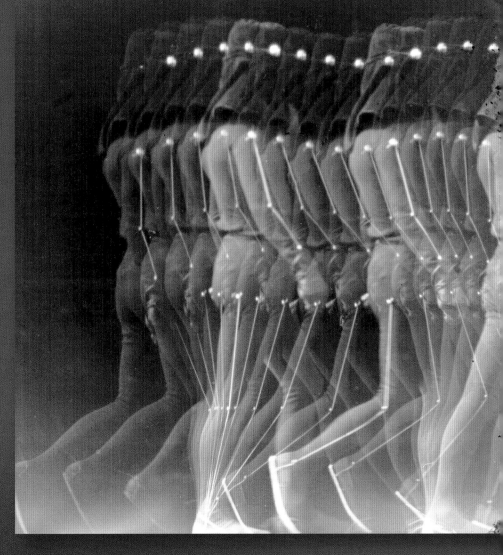

Chronophotograph
Marey shot these sequenced images (above) of the male body in motion in 1884. They are now in the collection of the Collège de France.

Man jumping
This sequence of pictures by Marey are preserved at France's Bibliothèque Nationale. During the Paris Olympics of 1900 Marey and his colleague Demenÿ took the opportunity to capture some of the best athletes of the day. They also photographed labourers at work.

EADWEARD MUYBRIDGE

In 1872 the pioneer British photographer Eadweard Muybridge received an odd commission from Leland Stanford, president of America's Central Pacific Railroad and the owner of large racing stables. Stanford wanted Muybridge to verify Étienne-Jules Marey's claim that a galloping horse sometimes had all four hooves off the ground at once. It took the eccentric photographer six years and the use of a dozen stop-action cameras synchronised in tight succession to complete the task. His conclusive images (right), taken at Stanford's stables at Palo Alto, California, were published in 1878.

THE PNEUMATIC TYRE – 1888

Air smooths out a bumpy ride

Thanks to the shock-absorbent qualities of rubber and air, the invention of inflatable tyres brought cyclists a new degree of comfort. The coming of automobiles upped production to an industrial scale and launched a constant search for improvements in what quickly became an indispensable item.

Puncture repair *Fixing tyres was a long and tricky operation until Édouard Michelin introduced removable versions (right).*

Trying his tyres *Born in Ayrshire, John Boyd Dunlop (left) studied at Edinburgh University to be a vet before moving to Ireland at the age of 27. There, 20 years later, he came up with the first pneumatic tyre.*

AHEAD OF HIS TIME?

In 1867 the Frenchmen Clément Ader, later a pioneer of aviation, had proposed fitting rubber to the iron rims of bicycle wheels. He started making his 'rubber velocipedes' the following year.

In the winter of 1887, a Scots-born Belfast vet named John Boyd Dunlop was distracted by a clattering noise from outside the house. His young son was tearing down the street on his new tricycle, jolted mercilessly by the cobbled paving stones. Glancing at the machine, Dunlop observed that the iron wheels were encircled only by thin ribbons of rubber that did little or nothing to cushion the impact. An idea came to him that he worked on over that winter until, in February 1888, he succeeded in covering the wheel rims with a tube made of fine sheets of rubber glued together at their extremities. He wrapped the tube in cotton and inflated it with a balloon pump. The pneumatic tyre was born.

Rival claims

In 1889 Dunlop gave a bicycle equipped with pneumatic tyres to a racing cyclist named William Hume, who rode it to victory in four successive races, helping to popularise the invention. That same year Dunlop founded a factory to make the tyres in Belfast. In 1891 he and a business associate opened another in Birmingham. At that point the heirs of an English inventor called Richard Thomson put in an appearance. Thomson, it turned out, had developed a pneumatic tyre 46 years earlier in 1845, but at that time there had been no obvious British patent for the invention and it had been forgotten. The matter went to court. Thomson's heirs won a judgment with regard to the authorship of the invention, but not over the way in which it was manufactured, which remained with Dunlop.

The removable tyre

In 1889 a French businessman named Édouard Michelin sought to help out an English tourist. The man's bicycle, fitted with Dunlop tyres, had broken down outside a factory producing agricultural machinery that Michelin ran near Clermont-Ferrand in central France. The Frenchman was much taken with Dunlop's invention, but he ended up spending long hours fixing the puncture; it took a whole night just to dry the inner tube onto the wheel rim. Nonetheless, Michelin saw that pneumatic tyres had a future and over the next few months he not only converted his factory to manufacture them, but also invented a removable version that was attached to the wheel rim by 17 screws. In future punctures would no longer be quite so much of a labour to fix.

From bicycle to automobile

The inflatable tyre proved a huge shot in the arm for the infant automobile industry. In 1895 the Éclair, designed by Amédée Bollée, was the first car to be equipped with Michelin tyres. That same year, in the celebrated road trial run from Paris to Bordeaux and back, the Éclair clearly demonstrated the superiority of pneumatic tyres over rubber strips.

Across the world firms now started manufacturing tyres to meet growing demand from motor-car manufacturers: in the USA Continental went into business in 1890 and Goodyear in 1898; Pirelli opened in Italy in 1899. With Dunlop and Michelin, these manufacturers between them gave tyres many of their familiar modern features, including grooves in the tread to improve road-holding (introduced by Continental in 1904) and the Bartlett Clincher rim, in-turned to hold the tyre in place, which was intitally introduced for bicycles in 1891 and adapted for cars by Michelin in 1917.

Handmade tyres
In 1916, when this picture was taken (left), tyres were still made by hand. The man seen here is putting the first layer of rubber around the rim of the wheel.

Radial plys and tubeless tyres

The next major step forward came in 1946 when Citroën front-wheel drive models were equipped with radial-ply tyres. The concept, patented by an American in 1915, was to strengthen the cross-ply design favoured at the time by most manufacturers with bands of cord set at 90° to the direction of travel. The radial design prevented the cords from rubbing against one another, reducing friction as the edges of the tyres flexed and thereby increasing both manoeuvrability and ride comfort.

Also in 1946, the American firm of Goodrich developed the tubeless tyre, fitted so tightly to the wheel rim as to do away with the need for an inner tube. Today the race is on to develop a fully efficient airless tyre, either with an inner lining strong enough to support the vehicle on its own or with the tread somehow linked to the hub by flexible spokes. In either case, the ultimate goal is to make Dunlop's invention puncture-proof.

TYRES ON TRAINS

In 1931 the Michelin company introduced rubber-tyred trains called Michelines that proved comfortable but unsuitable for carrying heavy loads. Similarly equipped underground trains went into service on the Paris Metro in 1956 and they are still in use on five of its lines.

Used-tyre mountain
Discarded tyres at a waste facility in Westley, central California. Fortunately, recycling tyres is relatively easy as there are several ways they can be re-used.

TYRE RECYCLING

What happens to the millions of used tyres that are discarded each year? Some get a fresh lease of life thanks to remoulding. Others go to cement works, where the rubber they contain makes them an efficient fuel. Increasing numbers are ground up and used in granular form to make synthetic sports surfaces. And some are buried to provide reinforced foundations for road construction.

THE ELECTRIC CHAIR – 1888

A new way of dealing out death

Seeking a more modern method of executing criminals than hanging, the authorities of the state of New York took a close interest in ongoing developments with electricity. The end result was the electric chair. Hailed as a more civilised means of execution, it was rapidly adopted by several states, but the claim that it was humane was far from proven.

The idea for the electric chair came from Dr Alfred P Southwick, a dentist from Buffalo, New York, who had seen a man electrocuted by accident. The unfortunate man had died quickly and apparently painlessly, and Southwick became instrumental in persuading New York State authorities that electrocution was the way forward for executions.

At the time, the 'war of currents' was raging between Thomas Edison, who backed direct current (DC), and Westinghouse Electric, which championed Nikola Tesla's alternating current (AC). George Westinghouse opposed the use of electrocution on humanitarian grounds, but Edison put research resources into it, assigning to the project two of his employees, Arthur E Kennelly and Harold P Brown – perhaps with ambiguous motivations. In his AC-DC battle with Westinghouse, Edison made much of the supposedly lethal nature of the AC system compared to DC, demonstrating the point by publicly electrocuting animals, including a circus elephant that had run amok. He may have thought that the use of AC electricity in the execution of criminals would discourage consumers from adopting it domestically.

New York authorities finally settled on the electric chair and passed the new form of execution into law on 1 January, 1889. Harold Brown was commissioned to construct the first models, powered by AC generators acquired second-hand. The method involved placing electrodes on the condemned person's head, with a damp sponge intended to help conduct the current. Despite evidence that death in the chair was far from quick and painless, it would be the principal means of execution in the USA until the 1980s, when it was largely replaced by lethal injection. It is still in use in some states.

Death by electrocution
A replica of the electric chair in use in the US state of Virginia in 1900. The leather straps were used to tie the condemned man down, securing his hands, torso and feet.

THE FIRST EXECUTION

Found guilty of killing his common-law wife with a hatchet and sentenced to death, 28-year-old William Kemmler was the first man to be executed by electrocution. His execution took place in 1890 before some 20 witnesses, and it quickly disabused those who thought the electric chair would provide a rapid and painless death. It took two separate discharges to kill him, inflicted a couple of minutes apart. The witnesses described distressing scenes, noting smoke coming from Kimmler's head along with a strong smell of burning. A subsequent autopsy revealed that the condemned man's brain had been 'cooked', and there were burns along the length of his backbone.

Photographic film 1889

The prolific American inventor George Eastman invented photographic film while seeking a replacement for the heavy glass plates that photographers previously had to use to capture images. In 1886 he came up with a length of paper covered with a photosensitive gelatin–silver bromide emulsion, wound onto a spool for insertion into the body of the camera. Two years later he used this to equip the first Kodak cameras, which for the first time brought photography within the reach of the average man or woman. In 1889 Eastman refined his invention by using transparent celluloid in place of the paper and film was born. As it was flexible, film proved ideal for cinema projection and within two years Edison invented the Kinetoscope. Soon photographic clubs and societies were sprouting in all the world's major cities. Industry, science and journalism happily abandoned the artists' sketches they had previously relied on and opted instead for the reliability and accuracy of photographs.

JUST ONE CLICK

Called Kodak because of Eastman's fondness for the letter 'k', the camera that he put on the market in 1888 was intended to be 'as easy to use as the pencil'. Handily portable and equipped with a 100-exposure film, the apparatus was designed to put photography within everyone's reach: Eastman's publicity slogan was 'You press the button, we do the rest'. Once all the photos had been taken, the camera went back to the factory for the film to be developed and a new one inserted; the whole device was then returned to its owner, ready for reuse.

The jukebox 1889

On 23 November, 1889, San Francisco's Palais Royale Saloon was jumping. The bar's owner had given Louis Glass and William S Arnold permission to install the first public phonograph. Invented 11 years earlier in the Edison stable, these ancestors of the gramophone could record and reproduce sound. Glass and Arnold's innovation was to add a slot into which coins could be inserted to start music playing. The machines were an instant success. Their golden age came in the 1940s and 50s, heyday of the great Wurlitzer jukeboxes, but they remained popular into the 1970s and beyond.

Music box
The Wurlitzer 850, known as the Peacock (left), was designed by Paul Fuller and introduced in 1941.

SCIENTIFIC MANAGEMENT
Factories run by stopwatches

In 1881 a Quaker engineer named Frederick Taylor introduced time-and-motion studies at a Pennsylvania steelworks. His initiative marked the beginning of the organisation of labour on rational lines, the principles of which Taylor would spell out in his 1911 monograph *The Principles of Scientific Management*.

Team photo
Some of the employees of Bethlehem Steel in Pennsylvania. Taylor found it easier to formulate his theories on productivity than to persuade men like these to act on them.

Born into Philadelphia high society, Frederick Winslow Taylor (1856–1915) was forced to abandon his law studies at Harvard because of failing eyesight. He radically changed his career plans, serving a double apprenticeship as a patternmaker and engineer at a Pennsylvania pump-manufacturing factory. He subsequently took a job at the Midvale steelworks, where he spent 12 years winning steady promotion, first to foreman, then research director and finally chief engineer.

Throughout his time at Midvale, Taylor tried to get the workforce to increase its productivity and was heaped with opprobrium for his pains, being variously called a tyrant, a slave-driver and a 'damned pig'. He left Midvale to work as a consulting engineer, notably with Bethlehem Steel, a firm he joined in 1898. That same year he helped to develop so-called 'high speed steel', an extra-hard alloy that could withstand high temperatures and that won him an international reputation. He finally gave up salaried work in 1901 to concentrate on publicising his ideas.

Increasing productivity

Taylor was convinced that the weak productivity achieved in factories was principally due to the amount of time that workers spent idling. To combat this tendency, he suggested a three-stage rationalisation process. The first step was to time each stage of production with a stopwatch to determine just how long should be allotted for its completion. The next task was to analyse individual tasks, breaking them down into their constituent parts and cutting out any superfluous actions. Finally, each worker had to be given precise instructions as to what he or she had to do. Taylor claimed that the elements that initially appeared most difficult to pin down, namely rest time and hold-ups, could in fact be calculated as precisely as everything else. To encourage the employees themselves to accept radical changes in their way of working, Taylor believed they should be offered the incentive of higher pay.

Gaining ground

The nascent trade unions, though, were not prepared to be pushed around. After several confrontations, Taylor was summoned before a congressional commission of enquiry, and in 1915 Congress actually forbade the use of stopwatches and piece-rate pay in military arsenals. Depressed and disheartened, Taylor

THE CASE OF WORKER SCHMIDT

To illustrate his theories, Taylor described an employee at Bethlehem Steel, to whom he gave the pseudonym 'Schmidt'. Schmidt was persuaded to shift 47 tonnes of pig iron instead of the normal 13 by simply following his supervisor's instructions: 'Now pick up a pig and walk. Now sit down and rest. Now walk – now rest.' In return Schmidt was offered $1.85 a day instead of the $1.15 he had earned previously. Taylor was aware that this approach was not suitable for everyone but he considered it appropriate for the likes of Schmidt, whom he described as 'so stupid and so phlegmatic that he more nearly resembles in his mental make-up the ox than any other type'. In fact, Taylor seems to have been stretching a point in this tale. In 1899 the labour force at Bethlehem Steel refused to change their working procedures, with the exception of one man, Henry Noll, who did agree to do so. On the one hand he endured the physical hardship of the extra work, on the other the abuse of fellow workers, and to cap it all he was made the prototype of the legendary Schmidt. But far from being the bovine character Taylor described, the real Schmidt jogged to and from work each day, served as a volunteer fireman and built a new home for his family in his spare time.

Organised workforce
Workers stand in rows at their posts in the AEG factory in Berlin around the year 1900. Some women employees can be seen on the left. Demand for women workers was increasing among employers because they were paid less than men.

Tied to the clock
The German director Fritz Lang brought home the negative aspects of the drive for productivity in his silent classic Metropolis (1926).

Women on the job
Female workers at the Midvale steelworks in Nicetown, Pennsylvania, in 1918. Midvale was one of the testing-grounds where Taylor first tried out his theories.

died of pneumonia shortly afterwards. Although he did not live to see it, Taylorism, as his theory of management became known, spread widely in the years after the First World War, especially in the USA, where industrialists were constantly on the look-out for increased efficiency to give them a competitive edge. Factories where foremen and multi-tasking workers had once rubbed shoulders began to change; specialised employees became the norm and a hierarchy of skills developed. Taylor's scientific management theories had opened the way for mass production and the widespread adoption of the assembly-line methods promoted by Henry Ford.

THE NEURON THEORY – 1889

Unveiling the brain's secrets

After long hours hunched over the microscope, taking meticulous notes of what he could see, Spain's Santiago Ramón y Cajal was able to reveal the overall architecture of the cerebral cortex and the inner structure of the cells that make it up. In doing so, he laid the foundations of modern neurophysiology.

In October 1889 Europe's most distinguished histologists – specialists in animal and plant tissue – gathered in Berlin for the annual conference of Germany's prestigious Society of Anatomists. The discipline's leading figures were all there. Yet it was an unknown 37-year-old, freshly arrived from Barcelona travelling in a third-class railway carriage, who was to leave his mark not just on the congress but also on our understanding of the nervous system. That man was Santiago Ramón y Cajal. He had spent an adventurous youth, first as a rebellious schoolboy and then as a medical officer with the Spanish army, before taking up an academic career. Now, opening his valise, he took out a microscope and a number of slides to begin his presentation entitled 'A Series of Preparations of Nerve Centres Made in accordance with the Golgi Method'.

In halting German, Cajal then invited the assembled scholars to look through the lens. Waiting for them there was a previously unknown world. Nerve cells could be seen with a clarity never before achieved. There were bulbous cellular bodies shaped like overfilled wineskins, dendrites in networks resembling tree roots, and long, isolated

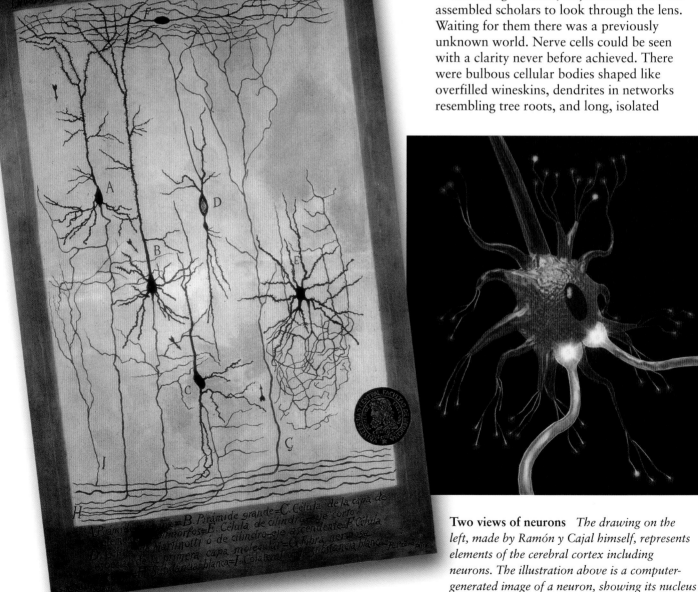

Two views of neurons *The drawing on the left, made by Ramón y Cajal himself, represents elements of the cerebral cortex including neurons. The illustration above is a computer-generated image of a neuron, showing its nucleus (the red dot in the centre) and the dendrites – filaments that act as receptor 'antennae'.*

A BREAKTHROUGH DISCOVERY

Camillo Golgi (1843-1926) studied medicine at the University of Pavia in Italy and later returned there as a professor of histology. No-one knows exactly when he discovered his breakthrough method of dyeing neurons with the aid of silver nitrate – Golgi himself always remained tight-lipped on the subject – but it is thought to have been when he worked as Chief Medical Officer at a hospital in nearby Abbiategrasso. From 1880 on Golgi's technique of selectively colouring neurons to make them easily visible under the microscope revolutionised the study of the nervous system, enabling Ramón y Cajal and others to elaborate the neuron theory. Some people, Ramón y Cajal among them, sought to play down Golgi's contribution by claiming that he had made his discovery by chance, when chemicals became accidentally mixed with nervous tissue on a work surface. So it was an irony of history when the two men were named joint winners, in 1906, of the Nobel prize for Physiology or Medicine, nominated in tandem for their 'work on the organisation of the nervous system'.

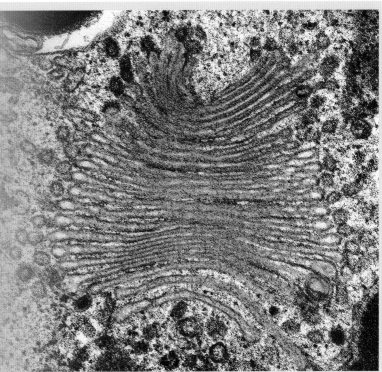

fibres – the axons – that seemed to have been thrown out like a boat's grappling hooks to establish connections with distant cells.

Burning the midnight oil

Ramón y Cajal had made his exhaustive observations over the course of entire nights spent in his austere laboratory at Barcelona University. In the course of this work, one crucial fact that the slides revealed to him was that the axons attached themselves to the surface of other cells but did not fuse with them. This seemingly innocuous discovery – which he spelled out in his *Manual of Histology* written the same year as the Berlin conference – constituted a fundamental step forward in our understanding of the nervous system. In effect, Cajal had concluded that 'the nervous system is organised in a discontinuous manner', with cells remaining independent of one another, separated by microscopic spaces that remained to be revealed. His hypothesis proved to be highly important. It became known as the neuron theory following the coining of the term 'neuron' by the German anatomist Heinrich Wilhelm de Waldeyer.

Conflicting views of the brain

Ramón y Cajal's intervention ended a debate that had divided the European scientific community for many years. Since the 1700s a consensus had grown up to the effect that there were two different types of tissue in the brain, described respectively as grey and white matter. Since Anton van Leeuwenhoek's early experiments with microscopes, people had also known that the nerves are composed of microscopic fibres organised in bundles. There was general agreement, too, about the overall structure of nerve cells, as described by the German neuro-anatomist Otto Deiters in a work published posthumously in 1865, detailing the cellular body, dendrites and a long, unique axon.

Yet one essential point remained obscure, dividing the opinions of specialist histologists. It concerned the exact nature of the network formed by the fibres. It was generally assumed that there were millions of fibres, if not more. For some histologists – the reticularists – the cells made up a diffuse, continuous web, like the waters of some vast river delta as seen

Golgi apparatus
Discovered by Camillo Golgi in 1898, the Golgi apparatus is found in most nucleus-bearing cells and is vital to the processing of proteins. The image (above) is a transmission electron micrograph, taken using a technique developed in the 1930s.

THE BEGINNINGS OF ELECTRO-ENCEPHALOGRAPHY

Following in the footsteps of the Italian Luigi Galvani, who had pioneered the study of electrical patterns in the nervous system, the Liverpool-based physician Richard Caton from the late 1870s on studied electrical impulses in the brains of living animals, principally rabbits. He demonstrated that when he stimulated a rabbit's retina with a beam of light, the electric current registered by an electrode implanted on the surface of the grey matter of the animal's brain underwent a marked variation. This discovery represented a first step in the discipline of electro-encephalography, initially a method of measuring the electrical activity of the cerebellum that would steadily advance through the 20th century, casting much new light on the role played by electrical impulses within the entire brain.

STUDYING EPILEPSY, HYSTERIA AND THE UNCONSCIOUS

While some scientists were investigating the structure and anatomy of the brain, others were trying to comprehend the relationship between consciousness, cerebral activity and certain forms of brain malfunction. At the Salpêtrière hospital in Paris Jean Martin Charcot, the founder of modern neurology, had been studying hysteria since the 1860s. Convinced that this personality disorder had organic causes, he carried out experiments on hypnotised patients that enabled him to bring to the surface evidence of ancient traumas that had long sunk below the surface of the

conscious mind. His demonstrations, some of which were held in public, attracted students and doctors from all over Europe. Sigmund Freud attended one in 1885 as he pursued his interest in the psychological origins of neuroses, a view that directly contradicted Charcot's own theories about them having organic causes.

Recovering from a hysterical fit
A female patient photographed at the Salpêtrière hospital in Paris in 1876. From a neurological point of view, epilepsy expresses itself as an electrical discharge in the brain.

MAKING CONNECTIONS

The word synapse was coined in 1897 by Sir Charles Scott Sherrington to designate the gaps between neurons where contact takes place.

from the air. But another group, the neuronists, believed instead that the nervous system was composed of independent units – the neurons. The debate went well beyond questions of pure science. As the home of the soul, the brain was considered by the reticularists to be something of a special case in the living world, a perfect creation incapable of being broken down into sub-sets. For the neuronists, on the other hand, the brain was just one organ among others, and therefore just like the rest it was made up of separate cells, even if these were highly specialised. Ramón y Cajal became the leading proponent of this school of thought.

A talented draughtsman

To facilitate his cellular observations, Ramón y Cajal used a method of dyeing neurons that had been developed by the Italian histologist, Camillo Golgi an arch-reticularist. So in a curious irony, he used Golgi's methods to refute Golgi's views. By exploiting the dyeing technique to the

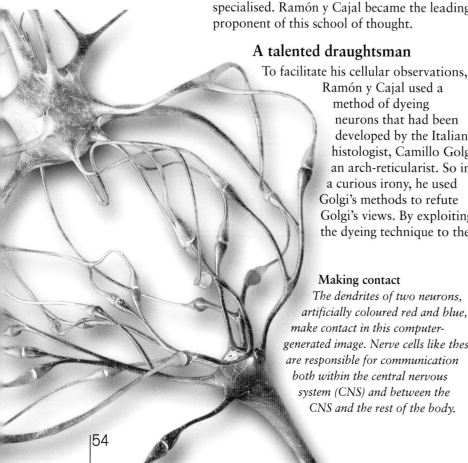

Making contact
The dendrites of two neurons, artificially coloured red and blue, make contact in this computer-generated image. Nerve cells like these are responsible for communication both within the central nervous system (CNS) and between the CNS and the rest of the body.

full, Cajal succeeded in penetrating the secrets of the nervous system, from its general architecture to its most intimate details. He observed cells in the shape of pyramids or rayed like stars, others 'with double dendritic bouquets' and 'short, ramified axons'. Thanks to his keen eye and a draughtsman's gift that he had exhibited since childhood, he was able to record the different categories of neurons in a series of drawings that were as precise in their details as photographs. Indeed, his drawings still serve as reference sources today. His results developed the work of the German neuropathologist Theodor Meynert, who had asserted in an 1884 book, *Psychiatry,* that the structure of the brain varies from region to region. The visual cortex, for instance, is made up of three layers, each dense with cells, while the primary motor cortex, which executes voluntary movements, is composed of a single type of giant, 'pyramidal' neuron.

Towards modern neurophysiology

Along with other scientists of his day, Ramón y Cajal laid the foundations of modern neurophysiology, even though at the time his theories remained largely unproven. It was only with the invention of the electron microscope in the 1950s that the neuron theory was definitively proved correct. It then became possible, for the first time, to see the synapses – the tiny spaces in which chemical communication between neurons takes place. The discovery led to a new vision of the brain as an organ of billions of neurons and almost infinite number of connections, through which it generates the thoughts, memories, actions and sensations that make up our daily lives.

The brassiere 1889

Since antiquity a range of devices have been used to support women's breasts in an attractive manner. For many generations the corset stiffened with whalebone did the job and was even considered a medical necessity, but over time critical voices were raised against it. In the 19th century a feminist and one-time political revolutionary named Herminie Cadolle invented a more comfortable arrangement that she called the *Bien-Etre* ('Wellbeing') because it allowed women to breathe more easily. The new garment resembled a corset cut in two below the chest, with the upper and lower sections held together at the back by laces. It failed to catch on, being cumbersome to wear, but it was exhibited at

Providing uplift
Marie Tucek's Breast Supporter (right) was a precursor of the underwire bra, but failed to catch on commercially.

Sensuous image
A publicity shot for a French bra manufacturer named Scandale (below). In 1937 they adopted the advertising slogan: 'Everything closest to a woman's heart bears the Scandale name'.

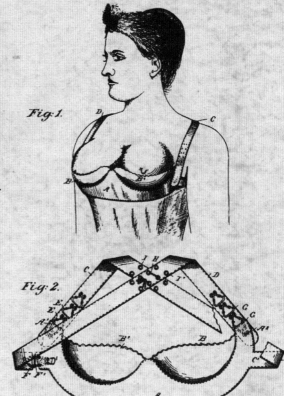

Scandale

HISTORY OF A WORD

Although the word 'brassiere' derives from French, the French themselves prefer the term *soutien-gorge* (literally, 'throat support') for the familiar undergarment, using *brassière* to refer to a type of corset. The earliest recorded English use of the word dates back to 1893, when a newspaper referred to women 'adopting for evening wear the six-inch straight boned band or *brassière*', and it first found its way into the Oxford English Dictionary in 1911. The abbreviation 'bra' became popular in the 1930s.

the Paris World's Fair of 1889, which gives it a claim to be the world's first brassiere.

In 1893 a New Yorker named Marie Tucek took out a patent for a 'Breast Supporter' – in effect, a brassiere with separate cups supported by shoulder straps and closed by hook-and-eye fastenings. Twenty years later a New York socialite called Mary Phelps Jacob – who later found fame as a patron of the arts under the name of Caresse Crosby – fashioned a softer garment out of two silk handkerchiefs, selling the patent a year later to the Warner Corset Co. At about the same time a fashion designer called Rosalind Kind came up with a different concept consisting of two triangles of fabric crossed before and behind.

It was only in the late 1920s that brassieres became standard wear. In the USA William Rosenthal, co-founder of Maidenform, began building them into dresses created by his wife Ida and co-partner Enid Bisset. When clients began requesting separate bras, they started giving them away with each dress. The bra allowed a freedom of movement that gave it a role in women's emancipation, but ironically by the 1970s it had become a symbol of gender stereotyping. The underwire bra, invented in the 1930s, became a commercial sensation reinvented as Wonderbra in the 1990s.

The electric oven 1889

Heat on demand
Made by the Carron Co in Scotland and installed in 1912, this cast-iron electric stove remained in use until 1970. It has two ovens, four burners (each with three separate heat settings) and an impressive, wood-mounted control panel.

Long before the beginning of recorded history, people were lighting fires to cook their food. The first known ovens date back to the Indus civilisation, which constructed open stone structures as early as 3200 BC.

The first recorded electric oven went into service at the Bernina hotel in Samedan, Switzerland, in 1889, where it was used to cook bread and pastries. Two years went by before a model first went on sale to the public, courtesy of the Carpenter Electric Heating Manufacturing Co of St Paul, Minnesota, in the USA. In 1896, William Hadaway filed a patent for an electric oven heated by a one-ring spiral coiled conductor. The ovens proved popular because they were not subject to the noxious leaks that plagued gas ovens at the time. Even so, they only became regular household features in the 1920s as the supply of electricity spread.

The electric kettle 1890

Cleaner and less bothersome than wood or charcoal and more reliable than gas, electrical energy was soon being applied to everyday household objects in the closing years of the 19th century. The kettle was no exception. In March 1890 Albert Gay and Robert Hammond took out British Patent No 4993 on a 'saucepan to contain liquid to be heated', which worked by induction. Their invention was an immediate success. By 1891 the firm of Crompton & Co was selling electric kettles in London, and in 1892 the Carpenter Electric Heating Co had a model available for the American market. The technology has continued to improve ever since. The first wireless kettles came out in the 1990s.

Cutaway prototype
Dating from the 1920s, this model was the first to have the element completely submerged.

A BRITISH INSTITUTION

Many of the improvements in kettle design were made in Britain, where electric kettles quickly became essential kitchen items. In 1892 electrodes replaced induction heating; they were later replaced in their turn by elements. In 1905 George James Gray patented a kettle device linked to an alarm clock. When the water boiled it flowed into a cup positioned on a balance that switched the electricity supply from the kettle to the alarm once the cup was heavy enough. The electric Teasmade was born.

Cities grow upward

In 1891 newspaper sellers in the streets of Boston, USA, had a new phrase to attract the attention of passers-by: 'How do you build a skyscraper? Read all about it in the *Daily News*.' The term was unfamiliar to all but a few sailors and yachting enthusiasts, for whom a skyscraper was a small, triangular flag at the top of a mast. Now applied to the ever-higher office and apartment blocks rising in Boston, New York and Chicago, the word had taken on an entirely new meaning.

Early high-rise
Chicago's Home Insurance Building (below) was built in 1884, but it did not become a 'skyscraper' until 1891 when the term was coined.

In 1870 the Equitable Life Assurance Society opened a new headquarters in New York. There was nothing particularly impressive about the five-storey office building, except that it was equipped with an Otis safety lift, provided with a mechanism to prevent the cabin from going into free fall if the cable broke. The coming of reliable lifts proved revolutionary: from that time on architects had free rein to build high, no longer constrained by the number of flights of stairs people could reasonably be expected to ascend. Even so, they were still limited by the amount of weight that load-bearing walls could support. Beyond eight stories, the walls became impracticably thick and window space was at a premium.

A new way of building

New methods were needed if buildings were to get taller. In Chicago a French-born architect and engineer named William Le Baron Jenney put his mind to the problem. He had plenty of opportunity to try out his ideas, for much of the town had been destroyed by the great fire of 1871. To rebuild it, high office buildings seemed ideal, offering large floor areas for the city's expanding businesses and providing the highest possible return for the developers who owned the plots on the ground.

For a time Jenney could see no solution to the problem of load-bearing walls, but then he remembered something he had noted in his youth. On a visit to Manila in the Philippines, he had been struck by the traditional houses, which rested on a frame of tree trunks with reed matting serving as partitions. The idea came to him that the solution to the problem of building high might be to rely not on the walls for support but rather on some inner framework that would bear the weight of the structure – one that was as tall as the building itself. He imagined an internal grid of beams and metal columns on which curtain-walls could be hung almost like reed mats.

Now determined to put his ideas into practice, he designed the Leiter Building in 1879. The seven storeys were supported on the outside by masonry pillars and on the inside by cast-iron uprights. The metallic framework was limited in its scope, but even so it proved sufficient to support a façade in which the dominant

Steel skeleton
The Carson, Pirie, Scott & Co Building, designed by Louis Sullivan, under construction in 1894. Known today as the Sullivan Center, it includes the School of the Art Institute of Chicago among its tenants.

feature was glass. The next step was to replace cast iron with a stronger metal, which turned out to be Bessemer steel, introduced in 1856 and soon being produced in large quantities. In 1884 Jenney built Chicago's Home Insurance Building, 10 storeys and 42m high, with a complete metal skeleton. A new style of building had been born.

Higher and higher

In 1892 two close associates of Jenney, Daniel Burnham and John W Root, built the Masonic Temple Building in Chicago, which at 21 storeys and 92m (302ft) tall was the first of a

succession of skyscrapers to claim the title of the world's tallest building. Two years later the 26-storey Manhattan Life Building in New York took the crown.

These early skyscrapers owed as much to progress in engineering as to architecture, only being made possible thanks to the technological advances of the day. On site the internal combustion engine, invented by Nikolaus Otto in 1876, gradually took over the heavy lifting jobs from human and horse labour. New welding techniques enabled workmen to attach metal girders without the need for any intervening metal connection. In the course of the 1890s Jenney and his School of Chicago colleagues (see above) developed a new way of laying foundations that involved sinking concrete caissons to spread the load on the pillars.

As showcases of progress, skyscrapers benefited internally from every state-of-the-art convenience, from lifts and electric lighting to steam central heating and running water dispensed from rooftop tanks fed by pumps from below ground level. From the early years of the 20th century on, first reinforced and then pre-stressed concrete provided cheaper alternatives as construction materials, although steel continued to be used in the USA, a major producer of the metal, into the 1960s.

Growing pains

In 1931 the Empire State Building was completed in New York, allowing visitors to gaze down on the city from the top of its 102 storeys and taking the world's tallest building crown at 380m (1,250ft). But with it the art of the skyscraper reached its apogee. By that time the Great Depression was temporarily thwarting developers' constant urge to build higher. For above all else, skyscrapers were temples celebrating the success of business, as witnessed by the 241m Woolworth Building (built 1913) and the 318m Chrysler Building (1930) among others.

After 1931 money for construction no longer flowed so freely. By that time, though, skyscrapers had already radically changed the appearance of America's cities, which had become vertical in orientation, radiating out

Space-saving shape
The Flatiron Building was designed to occupy a triangular corner plot in central Manhattan. Built in 1902 to plans drawn up by Daniel H Burnham, it currently houses offices.

Manhattan landmarks,
Built in the neo-Classical style, the 38-storey Equitable Building (right) dates from 1915. The glorious Chrysler Building (below), completed in 1930, was briefly the world's tallest until surpassed in the following year by the Empire State Building.

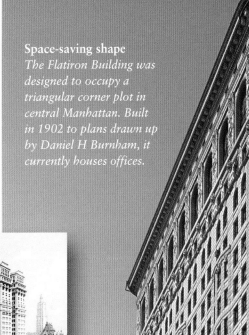

THE REVOLVING DOOR

Often used to give access to skyscrapers, revolving doors were invented in 1888 by an American of Swiss descent, Theophilus Van Kannel. Their great advantage lay in excluding draughts by providing an airlock between the interior of the building and the street outside. Doing so was not just a matter of comfort, for the difference in pressure between the ground and top floors of a skyscraper was so great that draughts, travelling up the lift-shafts, could become powerful enough to shatter upper-storey windows.

A COLOSSAL ENTERPRISE

Work on the Empire State Building began in March 1930, on foundations that had been laid 17m (55ft) deep. The figures involved in the building's construction are mind-numbing: 365,000 tonnes of materials, including 60,000 tonnes of steel and 6 million bricks, handled by a workforce of 3,000 men who between them clocked up 7 million working hours. Their ranks included the 'sky boys' who laboured without safety equipment on steel beams high above the void. Amazingly, none were numbered among the five fatalities recorded in the course of construction. Prefabricated sections helped to speed up the timetable; pillars, girders and 6,500 windows and frames were preassembled on site before being hoisted into place. In effect, the building site rose with the skyscraper itself, at a rate of four-and-a-half floors a week. The exterior was completed by November 1930, leaving only the internal fittings to be added. The building opened its doors to the public in May 1931.

from soaring business hyper-centres. In Europe architects proved more conservative and the public less supportive, which had the effect, with only rare exceptions, of delaying the appearance of really high buildings until the 1950s. They then started to make up for lost time: the following decade saw the London skyline radically altered, with the appearance of such landmarks as the 107m Shell Centre (1961), 117m Centre Point (1967) and the 177m Post Office – now BT – Tower.

In the USA, too, the 1960s saw a return to high-rise construction, fuelled by the booming

High-rise lunch *Construction workers on the RCA Building in New York take a break in 1932, perched some 250m above the ground (top).*

Going ape *Clutching Fay Wray in his mighty grasp, King Kong towers over Manhattan in a publicity image for the eponymous 1933 cinema classic. The film-makers used stop-motion animation and small-scale models to achieve their spectacular effects.*

economy of the day. Technological advances like tube-frame construction allowed buildings to exceed 400m, one of the earliest to do so being the World Trade Center (see box), which at 417m (1,368ft) overook the Empire State as the tallest building on the planet in 1972.

Skyscrapers have become a visible symbol of economic growth in many recently emergent economies. In Dubai, Taiwan and South Korea buildings are now being constructed that will approach or exceed the 800m mark. More than twice the height of the Empire State Building, these structures will literally scrape, if not the sky, then at least the clouds.

Under construction *Built in 1972, the twin towers symbolised US might much as the Empire State Building and the Statue of Liberty had done before them.*

THE TWIN TOWERS

The twin towers of the World Trade Center in New York, completed in 1972, rose 417m (1,368ft) high, an altitude long considered unattainable. Up to the 1960s skyscrapers more than 200m in height had stood up well to the wind, their worst enemy, thanks to a central concrete core supporting the traditional steel skeleton. But the higher they went, the larger the core had to be, thereby reducing the space available within the building. For the Twin Towers, architect Minoru Yamasaki opted for a revolutionary tube-frame design. The central core remained, but surrounded now by 47 load-bearing pillars that served to reduce its dimensions. Next came a series of 240 steel columns forming what was in effect a vast tube. The façade itself, composed of a rigid framework of tightly packed columns, also bore some of the weight of the structure. Until the attacks of 9/11 in 2001 the buildings were thought to be invulnerable, but the impact of the Boeing jets and even more the fires started by the thousands of tonnes of fuel the planes contained proved to be the towers' undoing.

Hong Kong cityscape
I M Pei's 369m Bank of China Tower (below left) was completed in 1989. César Pelli's 283m Cheung Kong Center (below right) dates from 10 years later.

GUSTAVE EIFFEL – 1832 TO 1923
Builder of much more than the Eiffel Tower

In some ways the Eiffel Tower has been Gustave Eiffel's bane. The most well-known landmark in Paris made his name famous all over the world, but it also obscured from view many of the other significant achievements of a man who was in some ways the archetype of the 19th-century civil engineer.

Engineer *extraordinaire*
A portrait of Gustave Eiffel taken by Stanislaus Walery early in the 20th century.

Inside view
An illustration of construction on the Bordeaux Bridge (below) from a magazine of August 1860, the month the bridge was completed.

If it had been built a few years earlier, Eiffel's greatest creation might have been called the Bönickhausen Tower, for that was his family name. On moving to France from the Eifel mountain region west of Cologne in Germany, Gustave's father changed his name for that of his birthplace, adding an extra 'f'. Gustave was born in the French city of Dijon in 1832 and grew up using both names, only abandoning the German part following the crushing defeat of France in the Franco-Prussian War in 1871.

Eiffel's parents ran a transportation business on the Burgundy Canal that made them prosperous but led them to neglect their largely unwanted son. So Gustave grew up in the care of his grandmother, and it was through his uncle, a local chemist and entrepreneur, that he had his first taste of the world of science and technology.

Fate intervenes

As a student Gustave studied chemistry at the Central School of Arts and Manufactures in Paris in preparation for a career at his uncle's vinegar works in Dijon. Then fate intervened in the form of a family quarrel, which ruled out that possibility shortly before he finished his studies. After graduating, he got a first glimpse of his true vocation at the Paris World's Fair of 1855, where he was much impressed by the imposing metal skeleton of the Palace of Industry, the first such structure to be built on such a scale in France. In time Eiffel's name would be indelibly linked with the golden age of iron and steel construction.

Thanks to an acquaintance of his mother, the young man got a job with a public-works entrepreneur named Charles Nepveu. In 1858, when Gustave was just 26, this man gave him his first substantial commission – to construct a bridge over the Garonne River in Bordeaux. Building a 500m-long structure was certainly a challenge, but he successfully completed the job in just two years. He also showed himself to be not just an effective manager but also a hands-on team leader: when one of his workmen nearly drowned, it was Eiffel himself who jumped in to rescue the man.

An astute businessman

The Bordeaux project made Eiffel's name. The Belgian Pauwels concern, which had bought out Nepveu, gave him other bridge-building projects in southwestern France. The next step for Eiffel was to set himself up as an independent construction engineer, teaming up with another young Central School graduate, Théophile Seyrig. Seyrig was rich, and their association worked very much to the advantage of Eiffel, who was already showing himself to be a shrewd businessman.

Eiffel & Co, as the new firm was called, at first specialised in building railway viaducts.

A RELIABLE EARNER

Although they were less spectacular than the great viaducts that Eiffel designed and built, prefabricated bridges provided a steady source of income for his company. Delivered in pieces along with instructions for their assembly, they were sold on every continent, providing both road and rail transport links. Eiffel's first customer was the French army, which used his products to span rivers in Indochina, where France was building a colonial empire.

At La Sioule and Neuvial, respectively 180m and 160m above ground level, Gustave honed the techniques that would serve him so well throughout his career. Yet it was a Portuguese project that put him in the top rank of his profession. The Maria Pia Bridge across the Douro River in Oporto traversed a valley 400m wide in a single span. At the time the biggest bridge in Europe, Eiffel built it in just two years, from 1875 to 1877. He then topped that feat by constructing the even greater Garabit Viaduct over the Truyère River in the mountainous Massif Central region of France, which opened in 1888.

In the intervening years the firm of Eiffel & Co had gone from strength to strength. In Paris they built the roof of the Bon Marché department store (1879) and the headquarters of the Crédit Lyonnais bank (1881). They were responsible for the inner framework of the Statue of Liberty (1884) and the dome of the Nice observatory (1886). Meanwhile Eiffel himself had to cope with several crises in his private life. First his parents died, and then in 1877 his wife, leaving him solely responsible for the upbringing of their five children. He took refuge in his work, where he was ably supported by his new engineering associates, Émile Nouguier and Maurice Koechlin. It was these two who first suggested to him the idea that was to become the Eiffel Tower.

Railway bridge
The Garabit Viaduct (above crosses the Truyère River outside the village of Ruynes-en-Margeride in the Cantal region of central France. Completed in 1884, it went into service four years later.

Spiral stair
Inside the Statue of Liberty, this spiral staircase takes visitors almost to the monument's full height of 46.5m. The statue was declared a UNESCO World Heritage Site in 1984.

THE STATUE OF LIBERTY – A GIFT FOR AMERICA

In 1878 the architect and sculptor Auguste Bartholdi turned to Eiffel & Co to provide the metal frame for the Statue of Liberty, which he was constructing as a gift for the American people to mark the centenary of the US Declaration of Independence. Eiffel conceived the structure as though it were the pile of some gigantic bridge, with four vertical uprights linked by cross struts. After it had been assembled, the frame was then dismantled for transportation by boat across the Atlantic, ready for the statue's inauguration on Liberty Island in New York Harbour on 28 October, 1886.

Masterpiece in metal
A photograph looking upwards from the foot of the Eiffel Tower (main picture) imparts the sheer quantity of metal that went into its construction. The tower was built over a period of 26 months from 1887 to 1889; two stages of its progress are shown here (right). Eiffel himself can be seen in top hat standing on a high platform of the structure (top right); his site manager and son-in-law, Adolphe Salles, poses on the spiral stair above him.

FIRM FOUNDATIONS FOR BRIDGES

To lay underwater foundations for the piles supporting bridges, Eiffel submerged open-ended sheet-metal caissons ballasted with concrete to make them sink into the riverbed. Compressed air was then pumped in to prevent seepage. Watertight shafts that were supplied by air tanks were also provided, giving workers access to the caissons to clear out debris and move tools and materials. Similar arrangements proved invaluable on the Eiffel Tower construction site to prevent the infiltration of water from the nearby River Seine.

A gigantic project

Plans were being drawn up in Paris at the time for a World's Fair to celebrate the centenary of the French Revolution of 1789. Some landmark project was needed as a focus for the festivities. At first Eiffel was unenthusiastic about the tower concept. Another architect, Stephen Sauvestre, did some preliminary work on the project before Eiffel bought back the rights, thereafter becoming the scheme's most enthusiastic promoter. Construction finally got underway in 1887. It took 225 workers labouring daily for 26 months to raise the 300m structure, which proved to be loved and hated in equal measure when it was finally completed. Ten thousand lamps lit it up for its official inauguration in 1889. Thanks largely to the Tower, the exhibition itself proved a great success with the public.

The tower made Eiffel himself rich and famous, but even in the year of its opening the sweet taste of success was snatched away. His reputation took a battering when the Panama Canal project failed, bankrupting many small investors. Eiffel, who had a lucrative contract to build ten metal locks for the waterway, found himself publicly pilloried. Worse was to come as he was put on trial for fraud and sentenced to two years' imprisonment. The judgment was annulled on appeal, but he was left embittered by the whole experience and retired from business.

Change of direction

Though in his 60s, Eiffel was still as alert and active as ever and embarked on a new career as a scientist using the tower to aid him. The upper stage of the Eiffel Tower proved an invaluable platform for astronomical and meteorological experiments. In 1898 Eugène Ducretet conducted trial radio broadcasts between the tower and the Pantheon. Cannily, Eiffel managed to preserve his masterpiece, due for demolition in 1910, by persuading the army that it had value as a telegraph mast.

Meanwhile, he was studying air resistance by dropping objects from the first level of the tower. Through his publications on the subject, he built up a reputation as an expert on aerodynamics and was rewarded with various awards, notably a gold medal from the Smithsonian Institution in Washington, DC. In particular he revealed the unequal forces operating on the inner and outer surfaces of aircraft wings, thereby adding to people's understanding of flight and the risks of stalling. In the years after 1910 various pioneer aeroplane manufacturers submitted models of their products to him for testing in his wind tunnel. The wind was, in a sense, a link between his early work on high structures and his later researches, and he was a passionate student of its effects to the end of his life.

THE WORLD'S MOST POWERFUL WIND TUNNEL

Eiffel had two wind tunnels built, the first at the foot of the Eiffel Tower and the second in the Parisian suburb of Auteuil. This last, which could test objects up to 2m in diameter against an airflow of 110km/h (70mph), was the most powerful in the world at the time. To achieve such speeds without having to use an impossibly large motor, Eiffel patented an ingenious system involving ducts and cones that proved highly profitable: aviation engineering firms from California to Moscow sought to buy up the rights to his invention.

Experimenting with wind *Based on Eiffel's device, this wind tunnel was constructed in 1914 at the Aerotechnical Institute at Saint-Cyr l'École, near Paris, where it was used to improve aircraft design.*

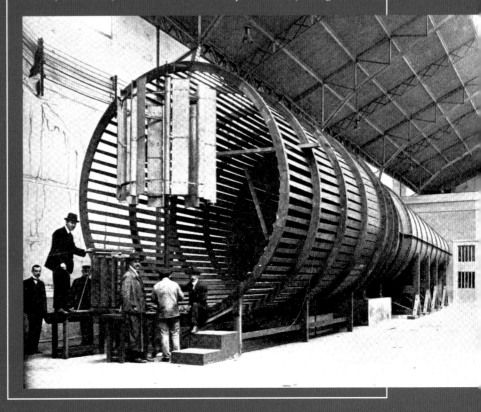

The zip fastener 1891

Elias Howe patented 'an automatic, continuous clothing closure' in 1851, but 40 years passed before his invention, now called a 'clasp locker', was put to use in a pair of shoes. These first fasteners had twin rows of hooks and grommets that were pulled together with a clasp. Unfortunately, they showed an unnerving tendency to open unexpectedly and were not a success.

More work was done by Gideon Sundback before the design was perfected in the USA in 1913. By that time small metal teeth had replaced the hooks, and the fasteners had a cloth backing. The name 'zip' or 'zipper' came from the B F Goodrich Company, which coined the term in 1923 when it marketed a line of rubber overshoes fastened with the device.

The escalator
1892

Although a first patent for 'revolving stairs' was issued to an American solicitor called Nathan Ames in 1859, the earliest known working model was built by a New York businessman named Jesse W Reno, who patented his own design in 1892. Four years later he exhibited his 'inclined elevator', which was powered by an electric motor, at Coney Island, where it attracted the attention of many thousands of passers-by. Four models were subsequently installed in department stores in New York and London.

At about the same time two other inventors, George Wheeler and Charles D Seeberger, were between them developing a rival system equipped with wooden steps. The Otis Elevator Company eventually bought out all three patents and marketed the device under the trademarked name of 'Escalator' in 1899. The steps themselves were fixed to chains driven by an electric motor via a system of sprocketed wheels and gears. The first city to install an escalator in its public-transport system was New York for an elevated-line station, followed by Paris in 1909 and Earl's Court station in London in 1911. Escalators with moving handrails were operational by 1920.

Escalators then and now
A design for a prototype escalator dating from 1894 (above) and a modern example (right) alongside stairs, typical of escalators in underground stations around the world.

THE MOVING WALKWAY

The first moving walkway was exhibited at the Chicago World's Fair in 1893. It ran in a loop along a lakefront pier to a casino and had two sections: in one passengers were seated, in the other they could stand or walk. The original design was improved over the ensuing years, and by 1900 the Paris World's Fair featured three separate electrically powered conveyors carrying passengers between the Champs Elysées and the Quai d'Orsay across the River Seine.

THE CAUSE OF MALARIA – 1892
Identifying the killer mosquito

Through dedicated and persistent work Ronald Ross, a Scottish physician posted to India, revealed that the parasite responsible for causing malaria is transmitted by a specific type of mosquito, the *Anopheles*. The discovery opened the way for research into ways of coping with the disease.

In 1881 the Cuban Carlos Juan Finlay discovered the role that mosquitoes play as vectors of disease in the course of his researches on yellow fever. Yet no-one at the time paid much attention to his findings. Recognition had to wait on the work of Dr Ross, a military physician posted to southern India. Ross first became interested in tropical diseases and their causes in the 1880s, soon after joining the Indian Medical Service. He took advantage of a brief return to Britain in 1889 to take a course in the emergent discipline of bacteriology. Thereafter he devoted himself increasingly to research on malaria, which at the time was cutting a swathe through the ranks of the Indian Army. The symptoms were all too familiar – high temperatures accompanied by shivering, vomiting and neurological complications. In the most serious cases the patient fell into a coma, generally dying within a few days.

For centuries past this mysterious malady had raged not just in the colonies but also in America and southern Europe. It got its name in the Middle Ages from the Italian *mala aria* or 'bad air', reflecting a belief that the disease was caused by miasmas given off by marshes. In the 19th century the condition was sometimes known as paludism, from *palus*, the Latin for 'marsh'.

A decisive encounter

An important preliminary step in identifying the cause of the disease had been taken in 1880, when the French military surgeon Alphonse Laveran had identified *Plasmodium* parasites in the blood of victims. Patiently, Ross set about trying to locate them under his microscope, but initially failed to do so. He was beginning to doubt if they really existed when he encountered Patrick Manson, another Scots doctor who had made his career in the colonies. Manson had established that

Airborne danger
A natural history illustration from 1918 shows the female Anopheles maculipennis, *one of over 100 separate* Anopheles *species responsible for transmitting malaria. This particular one is found in Europe and around the Mediterranean Basin into the Middle East.*

FAMOUS VICTIMS OF MALARIA

Malaria has been with us since antiquity and has claimed many millions of victims, including some famous ones. It is now accepted by most historians that Alexander the Great, who died at Babylon in June 323 BC after being struck down by fever for several days, was more likely killed by malaria than by poison. The disease was also responsible for the deaths of kings and Roman emperors. In the Middle Ages and the Renaissance years that followed, when 'marsh fever' was prevalent in Italy, it carried off a number of popes – 17 in the 13th century alone. Malaria has also been blamed for the deaths of Oliver Cromwell and Lord Byron, as well as the poet Dante and the painters Caravaggio and Raphael.

mosquitoes played a part in transmitting filariae, parasitic worms found in the human bloodstream, and he was convinced that the same must hold true for the agent responsible for malaria. After their meeting, Ross single-mindedly devoted himself to proving the 'mosquito theory'. In 1892 he demonstrated the presence of *Plasmodium* in the stomach of a mosquito that had drawn blood from a malarial patient. He then began experimenting on animals, placing healthy and infected birds together in a cage and finding that there was no cross-infection until he introduced mosquitoes, which caused the healthy birds to

NATURAL MAGIC

Used by Amerindians for centuries as a cure-all for fevers, cinchona was first imported into Europe in 1630 by missionaries returning from Peru, which brought it the name of 'Jesuits' bark'. It proved effective in both preventing and treating malaria, which at the time was widespread on the Continent. In about 1820 two French pharmacists, Pierre Pelletier and Joseph Caventou, succeeded in isolating its active ingredient, quinine. The molecule was synthesised in 1944, and derivatives like chloroquine were subsequently developed. In 1994 another anti-malarial substance, artemisinin, was derived from *Artemisia annua*, a plant used in Chinese medicine for more than 2,000 years.

Plant of many uses *Native to South America, Cinchona officinalis (right) is an evergreen shrub of the Rubiaceae family. Besides its medicinal uses, it is sometimes used as an ingredient in aperitifs and cosmetics.*

fall ill. The experiment indicated that the disease was transmitted through mosquitoes. To be more precise, through one particular genus of mosquito – the *Anopheles*.

A flood of discoveries

At approximately the same time three Italian scientists named Bignami, Grassi and Bastianelli were reaching similar conclusions. In pursuit of their researches, in 1898 they transmitted malaria to a human subject. They also established that it was the female mosquito that was the malarial vector.

Other researchers took up the torch, progressively casting light on the mysteries of the condition. Along the way they also identified several other infectious diseases carried by mosquitoes, among them dengue fever, filariasis and viral encephalitis. By the start of the 20th century three different species of *Plasmodium* had been identified: *P. falciparum*, *P. vivax* and *P. malariae*. Twenty years would pass before J W W Stephens, a Liverpool doctor, would identify a fourth type, *P. ovale*. It also took several decades to describe the life-cycle of the parasites in detail.

THE BIRTH OF TROPICAL MEDICINE

The world's first School of Tropical Medicine opened in Liverpool in 1898 thanks to a donation from a Liverpool shipowner, Alfred Lewis Jones. Ronald Ross taught there from 1902 to 1912. Patrick Manson, one of the founders of the study of tropical diseases, set up the London School of Tropical Medicine one year later, in 1899. Located at first in the Albert Dock Seamen's Hospital, it later amalgamated with the Hospital for Tropical Diseases in Euston. It still exists, now headquartered at Keppel Street in Bloomsbury.

Dangerous work *A biologist conducts experimental work in a high-security laboratory at the Bernhard-Nocht Institute for Tropical Medicine in Hamburg, the largest and longest-established in Germany.*

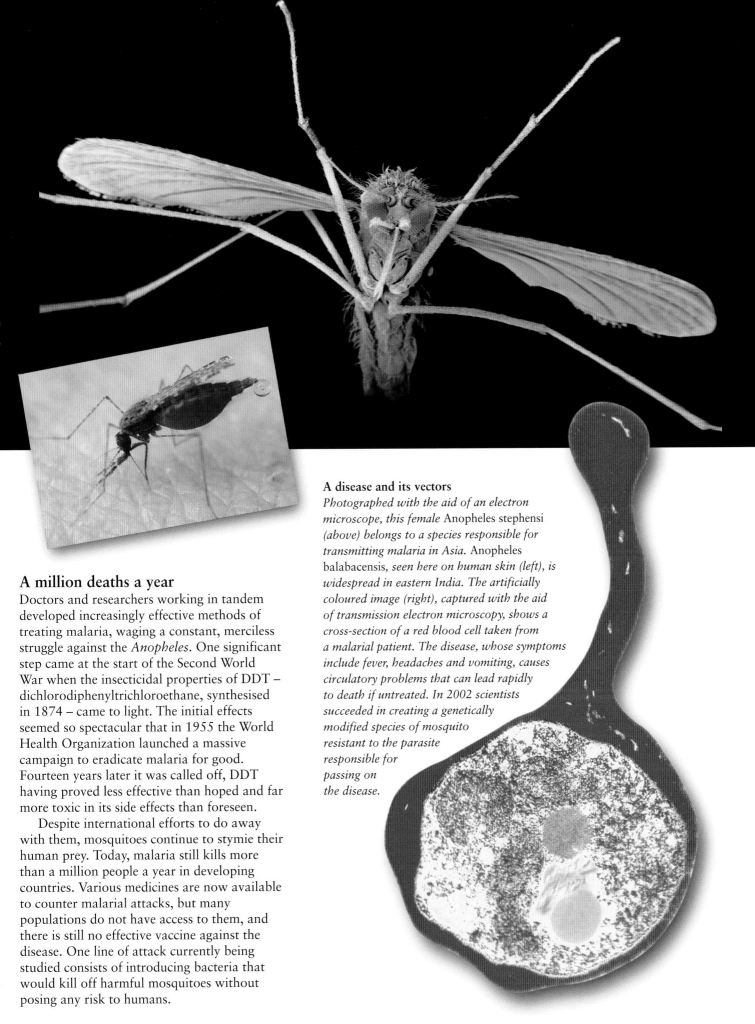

A million deaths a year

Doctors and researchers working in tandem developed increasingly effective methods of treating malaria, waging a constant, merciless struggle against the *Anopheles*. One significant step came at the start of the Second World War when the insecticidal properties of DDT – dichlorodiphenyltrichloroethane, synthesised in 1874 – came to light. The initial effects seemed so spectacular that in 1955 the World Health Organization launched a massive campaign to eradicate malaria for good. Fourteen years later it was called off, DDT having proved less effective than hoped and far more toxic in its side effects than foreseen.

Despite international efforts to do away with them, mosquitoes continue to stymie their human prey. Today, malaria still kills more than a million people a year in developing countries. Various medicines are now available to counter malarial attacks, but many populations do not have access to them, and there is still no effective vaccine against the disease. One line of attack currently being studied consists of introducing bacteria that would kill off harmful mosquitoes without posing any risk to humans.

A disease and its vectors
Photographed with the aid of an electron microscope, this female Anopheles stephensi *(above) belongs to a species responsible for transmitting malaria in Asia.* Anopheles balabacensis, *seen here on human skin (left), is widespread in eastern India. The artificially coloured image (right), captured with the aid of transmission electron microscopy, shows a cross-section of a red blood cell taken from a malarial patient. The disease, whose symptoms include fever, headaches and vomiting, causes circulatory problems that can lead rapidly to death if untreated. In 2002 scientists succeeded in creating a genetically modified species of mosquito resistant to the parasite responsible for passing on the disease.*

JULES VERNE – 1828 TO 1905
The father of science fiction

Extending to 62 novels and 18 shorter works, the vast literary output of Jules Verne has been enjoyed by, and left its mark on, generations of readers. This extraordinary legacy makes the visionary 19th-century author the outstanding pioneer of science-fiction writing.

Man of words and vision
A portrait of Jules Verne painted by Gustav Wertheimer (1847–1904), which now hangs in the Jules Verne Museum in Nantes.

At the time of his birth in 1828, nothing in Jules Verne's background seemed to predestine him to be the father of science fiction. His father was a solicitor who hoped to see his son follow in his footsteps. After a happy childhood in the city of Nantes in eastern France, Jules went to Paris at the age of 20 to study law, but spent more time reading literary classics than legal tomes. He passed long hours filling notebooks with comedies and dramas, and with the backing of Alexandre Dumas the Younger, author of *The Lady of the Camellias*, he had a light comedy called *Broken Straws* produced in June 1850.

A modest career as a dramatist seemed to be opening when Jules Verne met the explorer Jacques Arago, who introduced him to geography, astronomy, physics and chemistry. Captivated, Verne published two 'geographical' short stories in a wide-circulation Catholic magazine in 1851, prefiguring the themes of his later fiction. In 1857 he married Honorine Deviane, a young widow with two small daughters. To meet the expenses of family life, Verne took a job as a stockbroker while continuing with his literary efforts.

An inspirational publisher

Verne's writing career took a decisive turn for the better in 1862, when he met Pierre-Jules Hetzel, publisher of Balzac, Stendhal, Victor Hugo and George Sand. Hetzel clearly understood that the scientific discoveries of the time could provide rich material for a novelist willing to take modernity as his theme and express it in dramatic ways. He brought out Verne's *Five Weeks in a Balloon* in 1863. It proved

to be a masterstroke: 76,000 copies of the book were printed over the next four decades. And it was just the start of a long line of titles

With a publisher's eye for what the public. wanted, Hetzel sensed Verne's potential, signing him up in 1864 to write three books a year. Yet Verne himself was not convinced that he wanted to devote his talents exclusively to writing science-oriented adventure stories. Hetzel in fact turned down his second book, *Paris in the 20th Century*, which was only finally published in 1994. Set in 1960, it describes a capital criss-crossed by cable-car trains and private automobiles, and supplied with electric lighting (even for priests' monstrances), but where life is regulated by an omnipresent State and where the lust for money has triumphed over the love of culture. Hetzel found the work too pessimistic, noting in the margin of the text: 'My dear Verne, if you were a prophet, no-one today would believe your prophecies.' Yet despite some inaccuracies over details, many of Verne's ideas have proved surprisingly far-sighted.

Insatiable curiosity

Verne moved to Amiens in 1871, and set about providing himself with the information sources needed to enthuse readers about science and to take them on amazing imaginary journeys

---

A WEALTHY SUITOR

Several publishers had turned down Verne's first novel, *A Voyage in the Skies*, before he submitted it to Pierre-Jules Hetzel. While impressed by the originality, Hetzel found the construction clumsy and the writing only so-so. He also disliked the title and persuaded Verne to change it to *Five Weeks in a Balloon*. Verne recognised the importance of his new acquaintance immediately, confiding to friends, 'I'm going to hitch up with M. Hetzel, my wealthiest suitor'.

AN UNGENEROUS CONTRACT

In 1871 Verne signed his third contract with Hetzel, which paid him 1,000 francs a month (equivalent to about £3,500 today) in return for delivering two novels a year, a reduction from the three required previously. The deal was hardly generous, given the author's growing reputation and the fact that his novels were typically selling 30,000 or even 40,000 copies annually. But the two men had by then struck up a relationship that went way beyond the purely financial.

around the world and the cosmos. He joined scholarly societies, frequented libraries and read the works of all the best scientific popularisers. In addition, he subscribed to 10 or more daily newspapers as well as to every available scientific and travel journal, building up a file of thousands of note-cards arranged alphabetically by theme. In order to construct convincing future scenarios, he took care always to start from the existing state of technology and to achieve this he surrounded himself with advisers who could fill in the gaps in his own knowledge. These men included the

geographers Louis Vivien de Saint-Martin and Théophile Lavallée, the physiologist Louis Pierre Gratiolet and the chemist Henri Sainte-Claire Deville. When writing *From the Earth to the Moon* he asked his cousin Henri Garcet, a professor of mathematics at a prestigious Parisian school, to work out the calculations necessary to fire a projectile to the lunar surface. Later in his career he paid a mining engineer a fee of 2,500 francs to conduct a preparatory study for his novel *The Purchase of the North Pole*, which tells of artillerymen attached to the Gun Club of Baltimore who

Extraordinary adventures
Hetzel gave Verne's books distinctive, highly decorated covers from the start. The illustration above, by Léon Benett, is from Robur the Conqueror, *which was first published in 1886.*

Imaginary speedster
The Terror *(above), a combined high-speed automobile, aircraft, speedboat and submarine, sews panic across North America in* Master of the World *(1904).*

Real-life record-breaker
Jules Verne might have imagined a vehicle such as the experimental, rocket-propelled Blue Flame *(above). Fuelled with a mixture of hydrogen peroxide and liquefied natural gas,* Blue Flame *broke the world land speed record in 1970, reaching a top speed of 1,015km/h (630mph).*

a second satellite of the Earth besides the Moon, promoted at the time by a French astronomer named Frédéric Petit and floated in Verne's 1870 novel *Around the Moon*. On the other hand, although the subject interested him greatly, he kept out of the controversy about the possibility of life on Mars. In 1877 an Italian astronomer named Giovanni Schiparelli counted 79 straight lines on the surface of the planet that he took to be a network of natural channels, indicating the existence of seas and continents. The American Percival Lowell went further, interpreting the lines as canals forming part of some vast irrigation project and going so far as to construct a state-of-the-art observatory in Arizona to study them.

Similarly, Verne made no mention in his books of radioactivity, perhaps because it was discovered too late in his career to fire his imagination. Occasionally he was guilty of simple errors, for instance concerning the visibility of the solar eclipse of July 1860 (touched on in the 1873 novel *The Fur Country*) or the impossible orbit ascribed to the comet Gallia in *Hector Servadac* (1877). And enthralling though the plot of *Journey to the Centre of the Earth* may be, no-one will ever be able to penetrate the heart of the planet as Verne's heroes do.

set out to shift the Earth's axis in order to do away with the seasons so they can exploit the mineral riches of the polar region, to which they have bought the rights. Verne's knowledge became encyclopedic, and there were few regions of the world or fields of science, from astronomy to geology and oceanography, that failed to arouse his interest.

The odd aberration

Sometimes Verne espoused outlandish theories that were soon forgotten, like the existence of

A visionary at work

Verne himself repeatedly stressed that his main aim was not to prophesy future developments but rather to 'inform young people about the world in the most interesting possible way'. In fact he proved surprisingly accurate in many of the predictions that he made, and several of the discoveries and inventions that he described eventually materialised. At times

---

A DISCIPLINED LIFESTYLE

Jules Verne got up each morning at 5 o'clock, worked until 11 then ate a snack lunch followed by a siesta until 3 o'clock; he was in bed by 8 at night. Yet he was not just the armchair traveller that some of his biographers have claimed. As a young man he visited Scandinavia and Scotland. In 1867 he went to New York and Niagara Falls. He loved the sea and sailing and owned some splendid boats, notably the *Saint-Michel III*, a 31m steam-powered yacht .

AN EXPLOSIVE LAWSUIT

In 1896 Verne brought out *Facing the Flag*, telling the story of Thomas Roch, a misunderstood genius who invents a weapon called the Fulgurator powerful enough to blow up the planet. The book ends with Roch going mad and killing himself. The chemist Eugène Turpin, a real-life French inventor who had devised a new explosive, melinite, decided that Roch was based on him and started a libel action against the author. Defended by Raymond Poincaré, a future president of France, Verne won the case, even though he privately confessed in a letter to his brother that he had indeed drawn inspiration from Turpin's 'personality and actions'.

The joy of weightlessness
Zero gravity as enjoyed by the astronauts Ellen Baker and Michael McCulley on board the space shuttle Atlantis *(above), and as experienced by the three-man crew and their animal companions in the lunar projectile described in Jules Verne's* Around the Moon *(1870, top right). Another illustration from the same book (right) suggests how tiny Earth's inhabitants might look to imagined dwellers on the Sun if their size reflected the star's own dimensions.*

there was something almost eerie about his prescience. In *From the Earth to the Moon* and *Around the Moon*, he described humankind's first voyage to the Moon as taking off from Tampa, Florida – just 130km (80 miles) from Cape Canaveral, the actual launching pad for the Apollo programme from 1967 to 1972. Verne's vessel lands in the Pacific Ocean only 2km from the place where the Apollo 11 capsule splashed down at the end of its historic 1969 mission. In *Hector Servadac* Verne described the investigation of a comet by a space probe very much like the Rosetta, launched by the European Space Agency in 2004 and programmed to rendezvous with the

THE BIRTH OF SCIENCE FICTION

The term 'science fiction' was coined in 1929 by Hugo Gernsback, creator of *Amazing Stories* magazine; the magazine cover shown above dates from 1927. He intended the phrase to designate a branch of literature telling stories interlarded with scientific facts and futuristic visions that promised to waft readers far from their normal concerns through the intervention of extraterrestrial beings, marvellous worlds and unknown powers. The first science-fiction novel is often claimed to be Mary Shelley's *Frankenstein*, written in 1818. This new type of writing not only postulated potential future applications of science but also commented on the existing state of the world through metaphors that could be either comic or tragic, absurd or moralising.

In Britain, H G Wells published *The Time Machine* in 1895. Its hero travels 800,000 years into the future and discovers humankind divided between the Elois, pacific but effete descendants of the old ruling class, and the Morlocks, cannabilistic heirs to the proletariat who treat the Elois as food. In *The War of the Worlds* (1898) Wells describes warlike Martians devastating the Earth before themselves being wiped out by bacteria to which they have no immunity.

Although more renowned for his Sherlock Holmes adventures, Sir Arthur Conan Doyle also turned his hand to science fiction. In *The Lost World* (1912) he describes an expedition to the Amazon, where dinosaurs and other extinct creatures still exist, and in *When the World Screamed* (1928) Professor Challenger drills into the earth's crust where he awakens a giant creature that destroys his machine.

Science fiction really took off in the early 20th century with the 'pulps', magazines mostly produced in the USA devoted to stories on the imagined complications the future might bring.

A night at the opera
In this futuristic vision by Albert Robida (above), drawn in the late 19th century, opera-goers in the year 2000 return home in a bizarre variety of personal flying machines.

Frankenstein on film
Mary Shelley's novel was brought to the screen in 1931 by the American director James Whale, with Boris Karloff playing the monster. The film tells how Baron von Frankenstein fashions a being out of body parts, with the brain of a condemned murderer, and electrifies it into life thereby creating a monster. The Russian novelist Mikhail Bulgakov explored similar themes in Heart of a Dog (1925), in which a surgeon implants organs from an alcoholic petty criminal into a dog.

comet Churyumov-Gerasimenko in 2014. In *The Begum's Millions*, Verne described an artificial Earth satellite almost 80 years before the 1957 launch of Sputnik 1.

Verne also showed himself to be ahead of the game in *The Steam House* (1880), in which an armour-clad, steam-powered mechanical elephant, the brainchild of a Hindu prince, predicted the coming of modern bulldozers and tanks. The pilotless flying machines of *The Barsac Mission* inevitably call to mind today's drones. Captain Nemo's *Nautilus* in *Twenty Thousand Leagues under the Sea*, which ranges the world's oceans at will drawing its energy from the sea itself, preceded by almost 20 years the construction of the first electrically powered submarines, designed respectively by the Spaniard Isaac Peral and France's Henri Dupuy de Lôme and Gustave Zédé. These are just a few of the striking examples of foresight encountered in *Extraordinary Voyages*, the collective title that Hetzel gave to Verne's works. In his stories, the author also managed to reflect in his own unique way the great issues confronting the world at the end of the 19th century, among them colonialism, war between nations and the rise of capitalism.

Port-hole view
An illustration from Twenty Thousand Leagues Under the Sea *shows the fictional passengers on the* Nautilus *observing a giant octopus.*

Many of Verne's later novels suggest the concern he felt at the direction the world was taking and the dangers of the uncontrolled application of science. *The Begum's Millions* features an evil scientist planning the extermination of the weak. In *The Will of an Eccentric* Verne targeted pollution linked to the petroleum industry. In *The Village in the Treetops* he denounced the slaughter of elephants for their ivory. But his public proved cool to the pessimistic tone of these works.

One of the most extraordinary dream-merchants that literature has ever produced, Jules Verne died on 24 March, 1905, from complications associated with diabetes. The body of work he left behind still retains its fascination for readers around the world.

Early submarine
The Gymnote *(above), one of the world's first electric submarines, was launched in 1888. Jules Verne had published* Twenty Thousand Leagues under the Sea, *featuring the ocean-roaming* Nautilus, *18 years earlier.*

REMEMBERED IN SPACE

Verne was often unkind in his portrayal of astronomers, tending to depict them as difficult – the protagonists of his posthumous *The Chase of the Golden Meteor* are a case in point. But their real-life successors have not held it against him. One of the largest craters on the dark side of the Moon is called the Jules Verne and astronomers have also named three asteroids after him and his characters: Nemo, Verne and Nautilus were respectively discovered in 1951, 1988 and 1993.

Uncovering the secrets of viruses

Without knowing it, a Russian botanist named Dmitri Ivanovski first described a virus in 1892. It would take more than half a century for other scientists to elucidate the astonishing properties of these tiny agents of infection.

The flu virus
H1N1 (inset, right), often wrongly called the swine-flu virus, is actually the most common cause of human flu. A new strain derived from pigs was responsible for the 2009 pandemic, mutating to enable human-to-human transmission. It was described as a 'reassorted' virus because it contained elements of perhaps as many as five previously identified viruses in a new and dangerous combination.

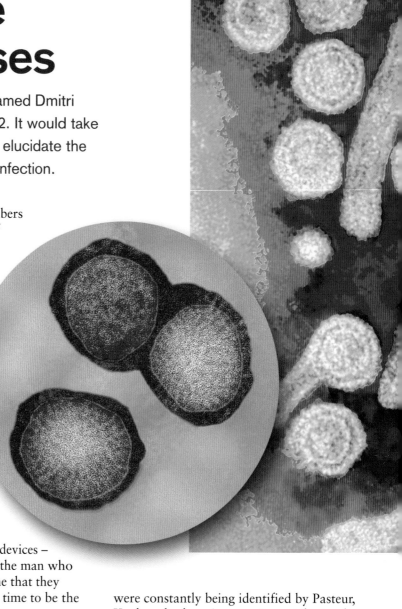

On 12 February, 1892, the members of the St Petersburg Academy of Sciences listened as a young Russian botanist, Dmitri Ivanovski, gave an account of his researches. For five years he had been studying diseases of tobacco plants, in particular tobacco mosaic, which had been ravaging plantations in the Crimea and Ukraine. In his address to the academy Ivanovski claimed to have transmitted mosaic from an infected to a healthy plant via diseased leaves, even though these had been passed through a porcelain water filter. This was surprising because the devices – called Chamberland filters after the man who invented them – had pores so fine that they retained bacteria, thought at the time to be the smallest infectious agents.

Confronted with this unexpected evidence of transmission, Ivanovski concluded that he must be dealing with unusually tiny bacteria, so small that he had not managed to observe them through his microscope. He also thought a bacteria-secreting toxin might be involved. He never suspected that he had flushed out the first known example of an entirely new and surprising family of germs: viruses.

Ivanovski's discovery later opened up a new field of biology, but it fell on sceptical ears and passed unnoticed at the time. In the last decade of the 19th century, bacteriology was all the rage and it was hard to imagine transmissible diseases being passed on by any other means. Staphylococci, streptococci, tubercular bacilli – new examples of bacteria responsible for human and animal infections

were constantly being identified by Pasteur, Koch and other pioneers. Bacterial microbes had two characteristic properties: they could be seen through the microscope and were retained by porcelain filters. Vaccines against viral infections like smallpox and rabies had been developed on a trial-and-error basis by Jenner and Pasteur among others, but the scientists working on them thought that they were dealing with highly specialised bacteria. Failing to find the agents responsible for rabies through his microscope, Pasteur had coined the term 'infrabacteria' to describe them.

A very special agent

In 1898 two German scientists, Friedrich Löffler and Paul Frosch, established that viral contagion also exists in the animal world. That same year Martinus Wilhem Beijerinck, a Dutch chemist, took a decisive step in

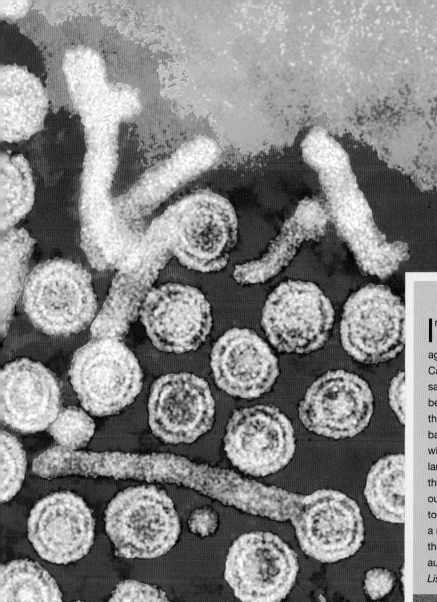

Liver-eaters
*The liver disease
known as hepatitis B
is caused by a DNA
virus belonging to
the Hepadnaviridae
family. Estimates
suggest that 350
million people
around the world
could be carriers.*

BENEFICIAL VIRUSES

In 1915 Frederick Twort, a bacteriology professor at the University of London, happened on a small agent that killed bacteria; two years later the Canadian Félix d'Herelle independently made the same discovery. Bacteriophages, as the entities became known, are viruses that infect bacteria and they were used from the 1920s on to combat bacterial infections. The treatments were prepared with the aid of micro-organisms that are found in large numbers in water and soil. With the arrival of the first antibiotics in the 1940s, phagotherapy fell out of favour, but in the past 20 years, as resistance to antibiotics has increased, researchers have taken a renewed interest in this therapeutic approach. At the same time bacteriophages have recently been authorised for use in the food industry to prevent *Listeria* infections.

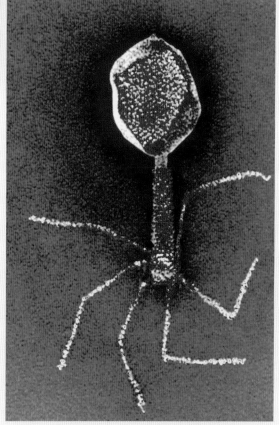

HOW TO RECOGNISE A VIRUS

Viruses contain a single nucleic acid, either DNA or RNA, but never both. They reproduce solely through the replication of this genetic material, being incapable of growing in size or of dividing in the way that living cells do. Viruses can only survive and multiply within living cells, whose DNA they assimilate. In effect, they are equipped to provide their own constituent parts, but cannot on their own fabricate all the proteins they need to survive. This definition of viruses, which still holds good today, was drawn up in 1953 by André Lwoff, a Nobel prize winner and one-time departmental head at the Institut Pasteur, who was one of the founders of molecular biology. Based as it is on their structure and behaviour, the definition applies to all viruses, whatever their host and the pathologies they induce.

Bacteria-eaters
*Bacteriophages are
viruses that only
infect bacteria.
So-called T-phages,
like this one (right),
attack E. coli
intestinal strains.*

disproving the prevailing bacterial hypothesis. Following up on the work of Adolph Meyer (a German who had been the first person to demonstrate the infectious nature of tobacco mosaic) and of Ivanovski, Beijerinck came to understand that the microbe responsible for the disease was a very unusual agent. It was alive in the sense that it could reproduce within growing tissue, yet 'fluid' since it could pass through filters. He called the organism a virus, adopting the Latin term for a poisonous liquid.

Other viruses were soon discovered that also had the capacity to traverse filters and to transmit infection from tissue to tissue. The agents responsible for such animal diseases

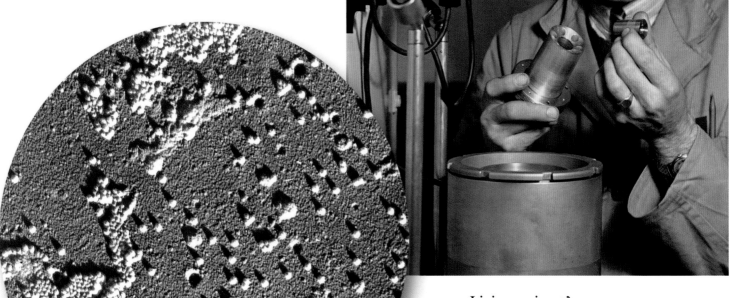

as foot-and-mouth in pigs and myxomatosis in rabbits were identified in 1898. Three years later Walter Reed, an American military physician, became the first person to isolate a virus responsible for a lethal human malady, yellow fever. Before long researchers were astonished to discover that viruses could infect all living things, including bacteria themselves. The first bacteriophages – literally, 'bacteria-eaters' – came to light during the First World War (see panel, page 77). Yet for want of any direct means of observing viral activity, a great deal remained unknown about the world they inhabited.

Polio virus
Taken in 1953, this image (left) confirmed the hypothesis that the agent responsible for causing polio was one of the smallest of all known viruses.

Microscope warrior
In 1950 a scientist prepares to use an electron microscope, only recently developed at the time, to observe a virus (top right). The new instrument magnified preparations 70,000 times.

Living or inert?

The scientific community remained divided over the reality of these miniscule entities that could neither be seen by the human eye nor cultivated using existing methods. In particular, researchers split over a crucial question: were viruses alive or inert?

In 1928 the British pathologist Arthur Edwin Boycott became the first person to give a plausible estimate of the size of viruses, putting them at 300 nanometres, which equates to 3 millionths of a millimetre. He also recorded some of their basic properties, but was unable to reach a conclusion on the living/inert question, so he qualified them as intermediary between the two. It seems he grasped the ambiguous nature of these micro-organisms, which appear to be living creatures when they penetrate cells and reproduce within them, but remain inert while outside cell walls waiting for the chance to infect a new host.

In the 1930s and 40s it finally became possible to observe viruses, thanks to the development of the electron microscope, and to cultivate them in cells. After decades of

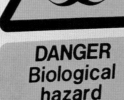

VIRUSES THAT CAN TRIGGER CANCER

In 1911 the American biologist Francis Peyton Rous, following on from the work of Germany's Ellerman and Bang and France's Amédée Borrel, demonstrated that a virus could trigger cancer in chickens. Subsequently other oncogenes (tumour-inducing viruses) have been identified in mammals. The first human oncogene was identified in 1957, when the Epstein-Barr virus (EBV) was found in African children infected with a rare form of jaw cancer. EBV is also involved in cancers of the nasopharynx, the area where the nasal passages meet the throat. The viruses that cause hepatitis B and C promote liver cancer, and some types of papillomavirus predispose women for cervical cancer, a fact that has permitted the development of vaccines. The eighth human herpes virus (HHV8) is associated with the rare skin cancer known as Kaposi's sarcoma, while the HTLV-1 virus, isolated in 1980, causes some forms of leukaemia. Like the HIV virus that causes AIDS, HTLV-1 is a retrovirus, a group that stores its nucleic acid in the form of RNA.

Security conscious
Increased concerns over biosecurity and the potential escape of dangerous viruses into the environment have led to a tightening of restrictions around research laboratories in recent years.

uncertainty, researchers at last had the tools they needed to elucidate the mysteries of the composition and replication of viruses.

A single nucleic acid

Virology became a recognised science in the early 1950s. Researchers showed that viruses are composed of a single nucleic acid (either DNA or RNA) in combination with proteins. As for virus replication, it became evident that this only takes place within a host's living cells. At the same time, cell culture opened the way for the discovery and characterisation of new viruses and also for the creation of viral vaccines for both humans and animals.

Today, much is understood about these omnipresent microbes, but they still pose a very real threat to health. New viruses show up regularly, as HIV did in the latter part of the 20th century, and their complexity and speed of evolution continues to outwit scientists. Even viruses that have long been known, like the flu virus, are in a constant state of mutation, requiring the continual development of fresh vaccines. And unlike bacterial infections, which can in the main be combated with antibiotics, only a small minority of viral infections respond to antiviral drugs.

International traveller
First isolated in 1983, the HIV virus responsible for AIDS (right) weakens the human immune system, rendering those infected susceptible to other, opportunistic infections. More than 30 million people have HIV worldwide.

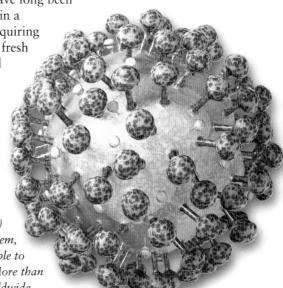

Making a more efficient motor

In 1878, working at the Technical University of Munich, Rudolf Diesel came up with the idea for an extraordinary machine. Based on the Carnot cycle proposed half a century earlier by Sadi Carnot, Diesel's new engine could in theory convert all its thermal energy into mechanical work. He spent the rest of his life designing engines that would approach this ideal.

THE FIRST DIESEL ENGINE

Installed in 1899 in a match factory at Kempten in Bavaria, Diesel's first engine had an output of 60 horsepower. After frequent breakdowns it had to be rebuilt from scratch.

Work on combustion engines had been ongoing since 1860, following the pioneering breakthrough of the Belgian engineer Étienne Lenoir, but Rudolf Diesel decided to take an entirely new approach. He wanted to create a radically different motor that would harness the very high temperatures and pressures associated with combustion to maximise output. His overriding concern was to modify the piston cycle within the cylinder in order to minimise the heat-loss. Unlike petrol engines, which required spark plugs to fire the fuel, Diesel's engine would compress the air in the cylinders so thoroughly that ignition would be triggered by the heat generated alone.

Diesel applied for a patent on his invention in 1892, but it took almost a year for the German authorities to accept his claims. At the time his idea hardly seemed revolutionary.

The engines designed by his fellow-German Nikolaus Otto had dominated the stage for more than a decade; Gottlieb Daimler and Karl Benz were already working on second-generation models. Diesel experienced every kind of trouble before he was able to sign a research contract with the Krupp engineering firm in 1893. Just four years later, he created an engine with 31.9 per cent efficiency, quite exceptional for the time. Initially intended to run on coal dust, in 1895 the engine was converted to diesel fuel, a less refined and therefore cheaper fuel than petrol.

A versatile tool

Production of the diesel engine under licence rapidly got underway in Europe, the USA and Russia. It was awarded the grand prize at the

Patent and prototype
One of Rudolf Diesel's first engines (right) and the patent he was eventually granted for his invention in 1892 (left). In the engine he created ignition was spontaneous, initiated by the heat generated within the cylinder, so obviating the need for spark plugs.

Turbo power
Invented in 1905, the turbocharger (above) is found in some but not all diesel engines. It links two turbines, using exhaust gases to increase output.

A SAD END

On 30 September, 1913, the steamer *Dresden* reached Harwich from Antwerp short of one passenger. Rudolf Diesel's body was retrieved from the sea some days later. While the exact circumstances of his death remain a mystery, all the evidence points to suicide. His finances were at a low ebb at the time, and he had been worn out by the legal and technical struggles associated with the development of his engine, which had already caused him to spend several spells in convalescent homes in 1898 and 1899.

Paris World's Fair of 1900. Initially there were numerous teething problems, including frequent leaks and breakdowns caused by the high temperatures and pressures involved, and it seemed that commercialisation of the engine might have got ahead of itself. Some industrialists lost interest but others, like Germany's MAN Group, persevered. In time the engine not only became reliable but was also found to be versatile and it was put to a variety of uses.

By the early 1900s the diesel engine was available in single or multi-cylinder forms. Used at first for machine-tools in factories and then for power stations, from 1903 it was adapted for naval use in ships and submarines. By the eve of the First World War, engines yielding 1,600hp were being built, and their popularity was growing because they were cheaper to run than petrol engines and less cumbersome than steam. From 1912 on they were put to use in locomotives, where by the 1930s they were increasingly replacing steam.

The first road vehicle to be equipped with a diesel engine was a truck built by the MAN Group in 1923, opening the way for their use in lorries capable of carrying loads of 20 tonnes or more. By the end of the decade they were becoming common in cars and buses, and by the end of the 20th century even some light aircraft were diesel-powered.

Increasing efficiency

The performance of the diesel engine improved after 1924, when the first injection pumps appeared. The Mercedes-Benz 260-D was introduced in 1936 as an experimental model, and by the 1970s diesel-powered cars were becoming common right across the automobile industry. At first the evil-smelling blue exhaust fumes that early models gave off proved a disincentive for many potential customers, but the direct-injection system developed by Fiat in 1996 finally solved the problem. By 2005 more than half the new cars registered across western Europe had diesel engines.

On the roof of the world
Opened in 2006, the Qinghai–Tibet railway (above) completes the rail link between Beijing and Lhasa. The trains are hauled by diesel engines, taking two days to cover a total distance of 4,064km (2,525 miles). The new section of line has claimed several world records, including highest station (Tanggula, at 5,068m) and highest tunnel (at 4,905m), as well as the overall altitude record for a railway at 5,072m in the Tanggula Pass.

The triumph of light

In 1900, when the Palace of Electricity was the centre-piece of the Paris World's Fair, electricity was seen as something magical, an inexhaustible clean energy that could drive machines, light streets and warm houses. It was personified as a fairy, a modern goddess missing only a temple dedicated to herself and her inventors. In 1937 that vision was reflected in a painting by Raoul Dufy, *La Fée Electricité*, commissioned for yet another World's Fair set in Paris.

In the closing years of the 19th century the power of electricity was finally mastered and tapped. Industry and transport were the first sectors to be transformed, but soon a whole range of new inventions and uses appeared as if by magic: the telegraph, telephone, gramophones and, above all, electric lightbulbs. Electric power spearheaded a revolution that shortened distances and contracted time in a way that would previously have been inconceivable. The new power source had its first moment of glory at the Paris World's Fair of 1900, when a statue of Electricity personified as a woman surmounted the metal tracery of the Palace of Electricity, showcasing a spectacular illuminated fountain. The exhibiton won Paris its nickname of *La Ville Lumière* ('The City of Light').

The spectacle fascinated visitors to the exhibition. One was the writer Émile Zola, who described in his novel *Travail* ('*Work*') an ideal society in which electricity served as a social bond: 'The day will come when everyone will have access to electricity as freely as to water or the wind ... It will circulate in the streets of cities as the lifeblood of society.'

The 1937 apotheosis

Over the ensuing decades Zola's prediction became reality. At the International Exhibition of 1937, electricity still roused passionate enthusiasm, even if by that time it had found its way into most people's homes. The Palace of Light and Electricity was erected in its honour on the terrace of the Trocadero, overlooking the Eiffel Tower across the River Seine. The building itself was sober in design, but at night it became a fairy palace, lit by immense solenoid coils that shot out bolts of light stretching for 7m. At its summit shone the lamp from the Ile d'Ouessant lighthouse, said to be the most powerful in the world. The exhibition also had an educational purpose; the official programme invited visitors to

Palace of light
A print by Fedor Hoffbauer shows the Palace of Electricity lit up at night (above). The 1900 World's Fair for which the palace was built attracted more than 50 million visitors.

Spirit of Electricity
A light bulb shines triumphantly in this advertisement for the AEG company of Berlin. Its founder, Emil Rathenau, bought the rights to produce electric light bulbs from Thomas Edison.

discover 'the social role of light' and its applications in homes and workplaces.

The exhibition organisers needed an artist with the requisite decorative flair to embellish the interior of the palace. They found the man they wanted in Raoul Dufy (1877–1953). His brief was to create a grandiose work glorifying electricity's inventors and their discoveries on a scale to match the building's spectacular special effects. The finished painting, *La Fée Electricité* ('The Fairy Electricity', see pages 84-87), curved in an arc around the spectators and featured 110 separate portraits of men of learning and science from ancient times to

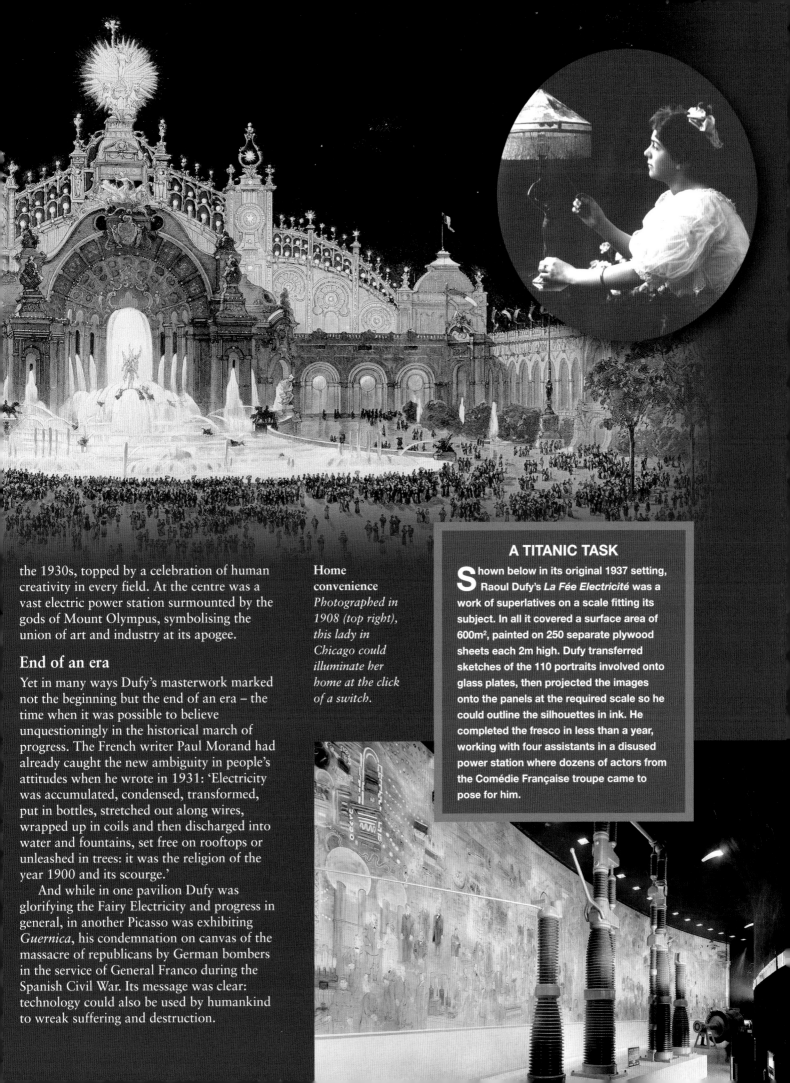

the 1930s, topped by a celebration of human creativity in every field. At the centre was a vast electric power station surmounted by the gods of Mount Olympus, symbolising the union of art and industry at its apogee.

End of an era

Yet in many ways Dufy's masterwork marked not the beginning but the end of an era – the time when it was possible to believe unquestioningly in the historical march of progress. The French writer Paul Morand had already caught the new ambiguity in people's attitudes when he wrote in 1931: 'Electricity was accumulated, condensed, transformed, put in bottles, stretched out along wires, wrapped up in coils and then discharged into water and fountains, set free on rooftops or unleashed in trees: it was the religion of the year 1900 and its scourge.'

And while in one pavilion Dufy was glorifying the Fairy Electricity and progress in general, in another Picasso was exhibiting *Guernica*, his condemnation on canvas of the massacre of republicans by German bombers in the service of General Franco during the Spanish Civil War. Its message was clear: technology could also be used by humankind to wreak suffering and destruction.

Home convenience
Photographed in 1908 (top right), this lady in Chicago could illuminate her home at the click of a switch.

A TITANIC TASK

Shown below in its original 1937 setting, Raoul Dufy's *La Fée Electricité* was a work of superlatives on a scale fitting its subject. In all it covered a surface area of 600m², painted on 250 separate plywood sheets each 2m high. Dufy transferred sketches of the 110 portraits involved onto glass plates, then projected the images onto the panels at the required scale so he could outline the silhouettes in ink. He completed the fresco in less than a year, working with four assistants in a disused power station where dozens of actors from the Comédie Française troupe came to pose for him.

La Fée Electricité

Dufy's monumental painting is now on show in the Musée d'Art Moderne de la Ville de Paris. It gathers together many of the greatest thinkers, scholars and inventors who contributed over the centuries to the discovery of electricity.

In the section reproduced here the characters include (from the right) Carl Gauss, the 'king of mathematicians', alongside Luigi Galvani, who tested the effects of electricity on dead frogs. Above him Henry Cavendish uses a torsion balance to estimate the average density of the Earth, while on his left Alessandro Volta poses alongside the electric battery he invented.

James Watt, who developed an improved steam engine, is presented under an image of a train emerging from the Gare St Lazare in Paris. The figure in uniform nearby is the physicist and military engineer Sadi Carnot, the founder of thermodynamics, while beside him François Arago explains his latest optical discoveries. Beneath Arago stands George Ohm, who established the relationship between electrical current, voltage and resistance. André-Marie Ampère, the founder of electrodynamics, occupies the place of honour at the foot of the power station. Above him stands Goethe, as if to establish the link between poetry and science.

La Fée Electricité (continued)
In this section of the fresco, cityscapes jostle with industrial scenes. The red patch on the ocean liner *Normandie* echoes that on Gramme's dynamo below it. Humphry Davy stands beside the battery he invented. Werner von Siemens, the inventor of the first electric passenger train, poses beneath a transformer.

The singular figure of Henri Becquerel, discoverer of radioactivity, seems lost in contemplation. Samuel Morse, the pioneer of telegraphy, turns his back on Émile Baudot, inventor of a rival code, seen here using his system to decipher a message. In the foreground, Wilhelm Röntgen stands behind the machine with which he discovered x-rays. On the far left two gigantic solenoids are the star features of a 14 July (France's Bastille Day) celebration, lit up by neon signs and fairy lights. Pierre and Marie Curie are shown involved in each other and their work. Finally Alexander Graham Bell and Thomas Edison, inventors of the telephone and electric light bulb, look off to the left to where an orchestra (out of view) pays homage to the Fairy of Electricity.

The private car transforms transport

The clack of hooves on cobblestones and the odour of the stableyard that pervaded city streets at the start of the 20th century soon gave way to the throb of engines and the smell of exhaust fumes as the century progressed. The motor industry was revving up.

In the last decades of the 19th century, the quickest, cheapest and most convenient way to get around town was the horse-drawn barouche. But ever since the start of the Industrial Revolution, machines had steadily been replacing animal as well as human labour. Steam engines had been employed in locomotives since the 1830s, but other motors, whether electrical or petrol-driven, seemed more promising for road transport, being easier to start and providing similar propulsive power in a much less cumbersome form. All three systems co-existed simultaneously until the first decade of the 20th century, when the petrol engine – powerful, compact, self-contained – established its dominance.

The first automobile
The three-wheeled Benz Motorwagon of 1885 (right) had a water-cooled single-cylinder engine with electric ignition and lighting.

Horseless carriage
The 1893 Benz Victoria (below) has been called the first mass-produced automobile. Benz sold 85 units that year.

The birth of the motor car

First invented by the Belgian engineer Étienne Lenoir in 1860, the internal combustion engine underwent rapid development in Germany in the 1880s. Karl Benz installed one on a tricycle in 1885, reaching speeds of 15km/h (9mph) that were impressive at the time. In the same year his compatriots Nikolaus Otto and Gottlieb Daimler tested a motor of their own on a bicycle equipped with stabilisers. It was Benz, though, who was responsible for the first true car, an elegant carriage he named the Victoria. Introduced in 1893, the vehicle was equipped with a multi-ratio gearbox involving a system of leather belts and pulleys of different sizes. The wooden wheels were chain-driven.

Strengths and weaknesses

The lessons learned in Germany spread across Europe. In 1890 Émile Levassor and Armand Peugeot began producing automobiles equipped with Daimler engines in France, laying the foundations of the French motor industry. In Britain the Santler brothers in Malvern and Frederick W Lanchester in Birmingham both produced innovative one-off models, but the

first production-line autos were made in 1896 by the Daimler Motor Company of Coventry, working under licence from Gottlieb Daimler. These early cars could already reach a speed of 15mph (25km/h), but few people could afford one: the price tag was equivalent to 10 years' salary for a factory worker, so motoring was mainly a hobby for the well-to-do. Drivers had to fire the burners with a lighter, then vigorously turn a starting handle to start the engine. Repeated breakdowns required them to get their hands dirty.

Cars were expensive to run, evil-smelling and dangerous. They inherited their brakes from horse carriages – a single steel band with brake shoes applied to the rear wheels. Even at 15mph, they were incapable of bringing the vehicle to a sudden stop – a failing that in 1896 would cost Émile Levassor his life. Experiments with disc-style brakes, which use friction to slow or stop wheels, began in England in the 1890s and were patented by Lanchester in 1902, but took another half century to be widely adopted. Even so, cars had some advantages over horse-drawn vehicles. They went faster, took up less space and did not need feeding when they were not in operation. They also produced no manure, perhaps a drawback in the country but a plus

REGISTERING VEHICLES

In the days of steam carriages, draconian regulations imposed a maximum speed of 4mph on British roads and insisted that a man walking with a red flag should precede the vehicles. By the 1890s, with the first petrol-driven cars, the rules were eased: vehicles weighing under 3 tonnes no longer required a flag, and the speed limit was raised to a heady 14mph. By 1903, as cars became more common, some means of identifying vehicles was needed. The Motor Car Act of that year stated that all vehicles in Britain had to be registered with the appropriate local authority and the registration marks displayed. Vehicle registration was not centralised until 1930.

in towns. The title of a magazine first published in the USA in 1895 was prophetic; claiming to be the world's first automotive magazine, it was called *The Horseless Age*.

The start of the motor age
From the early days of cars, manufacturers entered their vehicles for races and endurance

Unplanned stop
Breakdowns were frequent on early car journeys. This photograph (above) was taken in France in the last decade of the 19th century.

DRIVING LICENCES

The 1903 Motor Car Act introduced vehicle registration and required drivers to be licensed. The minimum age was set at 17, but there was no driving test (compulsory testing of drivers was not introduced until 1935). The act did away with speed limits, enacting instead an offence of dangerous driving, punishable by a fine of up to £20 for a first offence and £50 or three months' imprisonment for a second. The penalties were first enforced at Bow Street Magistrates Court in January 1904 against a motorist accused of being drunk in charge of a vehicle and failing to produce his licence; he was fined £10 for the first offence and 5 shillings for the second.

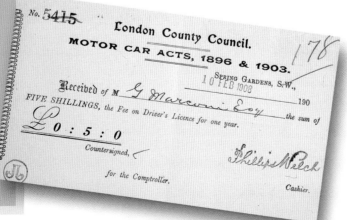

Pioneer's permit
A receipt acknowledging payment for a driving licence (left). The name on the receipt is Guglielmo Marconi (1874–1937), pioneer of wireless telegraphy.

From private phaeton to people's car
Two women enjoy an outing in a petrol-powered Panhard–Levassor(left); built in 1896, it had a top speed of 20km/h (12.5mph). The Model T Ford (above) was first produced in 1908, using production-line methods that dramatically brought down the price of cars, revolutionising the industry as a whole.

THE PETROL PUMP

Early cars were thirsty for fuel, compelling drivers to carry with them ample supplies of bottles of petrol, which could be bought in hardware stores. In 1898 John J Tokheim, a Norwegian immigrant to the USA who owned one such store, decided for reasons of security to dig an underground petrol reservoir outside his premises. He then improvised a pump to draw up the fuel and added a counter to measure the flow. In effect he had invented the petrol pump, and he duly patented his invention in 1901.

PUT OUT TO GRASS

By 1910 motor taxis already outnumbered horse-drawn cabs in London by 6,300 to 5,000. By the 1920s their victory was complete and horses were rarely seen on the capital's streets.

trials that quickly caught the public's imagination. Cars stopped being mechanical curiosities and started to become fascinating and desirable objects. In 1899 *Scientific American* recorded that there were already 6,546 cars in France, 688 in the USA and almost 500 apiece in Britain, Germany and Austria. Over the ensuing decade motoring was to become much more affordable, thanks largely to the efforts of Henry Ford. The Ford Model T cost only $650 in 1910 and was particularly popular with farmers, for whom it offered a lifeline from rural isolation. The number of firms operating in the motor business surged: by the turn of the century there were more than 600 in France, at least

190 in the USA, 110 in England and 80 in Germany. In cities automobiles now occupied the crown of the road, relegating horse-drawn vehicles to the gutters.

From 1904 on, cars also started to become more comfortable to ride in. Chariots open to the winds gave way to enclosed vehicles in which the bodywork protected the occupants from foul weather and the dust of the roadway. Only the wealthy could afford these luxury automobiles, whose interiors sometimes resembled ladies' boudoirs. Even so, the new motors, which could go farther and faster than ever before, were starting to change the look of the countryside. Cars required an extensive, well-maintained road network, spurring fresh

THE TRIUMPH OF PETROL

Motor manufacturers continued to produce electric cars until 1909, but from that time on petrol-powered vehicles, lighter and more self-contained, reigned supreme.

enthusiasm for the surfacing techniques devised by John Loudon McAdam in the early 19th century. To help motorists find their way, automobile clubs and private enterprise came together to provide maps, road signs and signposts. The Automobile Club of Great Britain was founded in 1897, becoming the Royal Automobile Club in 1907 thanks to the patronage of King Edward VII, himself an enthusiastic motorist.

The spread of motoring stimulated growth in sectors such as plastics, rubber, machine tools, steel and the glass and textile industries, as well as in service sectors such as insurance and consumer credit. In time, its development would become the symbol of a nation's success and a marker of economic health.

Electric record-breaker
Designed by Camille Jenatzy, a Belgian engineer, the electrically-powered Jamais Contente *('Never Satisfied', below), was the first vehicle to exceed 100km/h, reaching 105km/h (65mph) in a timed trial in 1899.*

The automobile

Powered first by steam, briefly by electricity and then by petrol, cars were the end product of a long succession of technical advances. Development has continued as engineers and designers strive to overcome the challenges of speed, comfort and road safety, inspiring multiple innovations.

UNDER THE BONNET
ENGINES AND TRANSMISSION

The first self-propelled automobiles dated from the early 19th century and were steam-powered. Petrol motors were introduced in the 1880s, and soon proved more popular than their electric rivals. Diesel engines were available from the 1890s, but diesel-powered cars did not become popular until the 1970s when the rising cost of fuel began to make efficiency more important. At first a starter handle was needed to start the engine; electric starters were introduced in 1912. Early automobiles had a system of belts, pulleys and cogs to modulate the rotational speed transmitted from the engine to the wheels. The first gear-boxes were fitted in Panhards and Mercedes vehicles at the turn of the 20th century. Daimler brought in semi-automatic gears In the 1930s; fully automatic systems appeared just after World War Two.

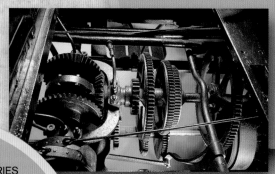

A Daimler petrol engine in an 1894 Panhard–Levassor model (right).

LIGHTING
HEADLAMPS AND BATTERIES

In the very early days of motoring, the only lights available to drivers were acetylene lamps similar to those used in mines. In 1881 a French inventor named Camille Faure developed a lead-acid battery (invented 22 years earlier by Gaston Planté) to power the motor and this provided the energy necessary for a variety of appliances, from windscreen wipers to electric windows. The first electric headlamps came in 1898, but it took a decade for them to catch on. By law, cars had to be provided with lights from 1919 on.

Early headlamps were powered by gas or acetylene (above).

BRAKES
FROM BLOCKS TO ANTI-LOCK

As early as 1550 wagons in European mines were equipped with levers attached to wooden chocks that could be braced against the wheels to slow their rotation. The same system worked for horse-drawn vehicles, but proved inadequate for cars with their higher speeds. In 1902 Renault introduced drum brakes, involving a system of shoes and pads that pressed against the inner surface of a revolving drum connected to the wheel. Disc brakes were patented by Frederick Lanchester in the same year, but only came into common use in the 1960s. In the 1920s a hydraulic mechanism replaced the steel cables previously used to operate brakes. Anti-lock braking and emergency brake assist completed the picture in the later years of the 20th century.

ROAD-HOLDING
WHEELS AND TYRES

The first cars borrowed their wheels from horse-drawn cabs and barouches. They were made of wood and rimmed with steel. The idea of covering metal-rimmed wheels with strips of rubber was first used on bicycles. After 1888 flat rubber bands were replaced by pneumatic tyres filled with air to absorb shocks, as first developed by John Boyd Dunlop. Two years later the Michelin brothers invented the removable tyre, which was also soon put to use on cars.

COMFORT
WINDSCREENS AND WIPERS

Glass windscreens first appeared in 1899, but they had a tendency to shatter until laminated glass, which cracked but did not break in pieces, was introduced. Invented in 1903 – the same year that the windscreen wiper first appeared – laminated glass only became available in mass-production models after 1929. The rear-view mirror dates from 1896.

An early steering column (above).

STEERING
FROM THE TILLER TO THE WHEEL

In the 15th century horse-drawn vehicles could change direction with the aid of a peg attached to the front axle. This fragile arrangement was replaced in the 19th century by a rack-and-pinion system. Early cars were equipped with a metal shaft linked to the steering rod. Steering wheels first appeared in 1898 in Panhard–Levassor cars. Power steering dates from 1950.

The steering wheel and horn of a four-cylinder Daimler (below).

THE CAR OF TOMORROW
ECONOMICAL AND NON-POLLUTING

Car designers had been quick to adopt the combustion engine, but the oil shocks of the 1970s forced a review of fuel consumption. Diesel engines benefited as their thermodynamic efficiency made them more fuel efficient. The quest became more urgent as oil supplies shrank and the environmental impact of fossil fuels became evident. Engines designed to run on LPG (liquefied petroleum gas) were introduced in 1979, and by the turn of the century hybrid vehicles combining electric motors and combustion engines were putting in an appearance. Yet for all the research put into developing more efficient engines, cleaner fuels and electric cars, the non-polluting model has not yet been invented.

The Fiat 509 Torpedo (above).

Suspension and wheels on the electrically powered Jamais Contente (below).

SUSPENSION
ENSURING A SMOOTH RIDE

The only suspension in use on horse-drawn vehicles in the 16th century was a metal bar supported by the wheel hubs. The next step was to suspend the body of the carriage on chains or leather straps across the axles. By the late 17th century these were replaced by springs composed of stacked metal strips, but the strips proved susceptible to rust. The French car manufacturer Mors adopted coiled shock absorbers in 1903. Thirty years later Citroën cars were equipped with torsion bars that adapted to bumps and jolts. In 1925 Georges Messier invented pneumatic suspension, using the elasticity of gases to absorb shocks.

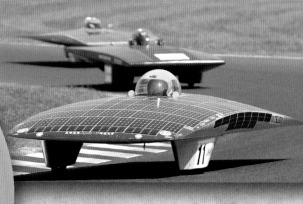

Solar-powered Taiwanese Apollo-5 experimental vehicles, built for racing (above).

A popular new art form is born

Two decades elapsed between the first public screening of moving pictures and the time when cinemas became standard features of all major European cities. The turn of the century saw the dream industry gradually taking shape.

A new view of the world

The Lumière brothers' Cinematograph (below) was both a cine-camera and a projector, providing all the equipment needed to make and show early film loops like A Train Arrives at La Ciotat Station *(right). A rival system called the Kinetoscope emerged from the Thomas Edison inventions stable; these frames (left) are from an experimental Kinetoscope sound loop called* Dickson Violin *after the machine's inventor, William Dickson. The drawback with Kinetoscope was that it only allowed films to be seen by one person at a time through a peephole in the viewing cabinet.*

The train loomed ever closer, showing no sign of stopping. As the cinematic image seemed to crash through the screen and hurtle straight into the hall, a few spectators in the audience actually screamed. Then the action was over and the lights went up. The date was Saturday, 28 December, 1895. The place was the Indian Room of the Grand Café on Paris's Boulevard des Capucines and the event was the first ever public showing of 'cinematographic views'. Not surprisingly, filmed images of people and objects moving just as they would in real life came as something of a shock, but every one of the 33 individuals who had paid the 1 franc entrance fee to see them considered the money eminently well spent.

They had been attracted by a notice pinned up outside the hall announcing the arrival of the Lumière Cinematograph invented by MM Auguste and Louis Lumière. 'This device …', the poster promised, 'records all the movements that take place before a lens over a given period of time as a series of instantaneous prints and then projects them in the form of life-size images on a screen before an audience.' The programme was 10 short films, each about a minute long, showing scenes from everyday life, including such titles as: *Workers Leaving the Lumière Factory, A Train Arrives at La Ciotat Station, Baby's Breakfast, Carriages in Motion, The Sprinkler Sprinkled* (this last a comic effort showing a gardener getting doused with his own hose).

continuous motion when the disc was spun; a later, more sophisticated version, known as the zootrope, appeared in 1834. A year earlier the kinetiscope had combined the principles of the revolving disc and the magic lantern to cast images onto a screen. Next in line was Charles-Emile Reynaud's praxinoscope of 1876, which consisted of a drum covered with mirrors placed inside a larger drum with a strip of images lining its inner surface; when the drum was turned the images were reflected from the mirrors and appeared to move.

Another crucial contribution came from Étienne Jules Marey and Eadweard Muybridge, pioneer photographers of motion whose researches cast invaluable light on the problems of capturing and storing images of objects in motion. The first person to devise a system truly capable of achieving that goal was William Dickson, an employee of Thomas Edison. His Kinetoscope took a succession of shots on standard Eastman Kodak photo stock that were then glued onto a transparent celluloid strip 35mm in diameter, with perforations along its edges. The device captured motion that could be viewed, but it did not permit communal viewing; only one person at a time could watch the short films unroll, hunched over the eyepiece of the small cabinet that contained them.

Combining camera and projector

The Lumière brothers' Cinematograph improved on the Kinetoscope by fixing the images of people or objects in motion on a reel of film that unspooled behind a lens, thus capturing a series of images at very short intervals. The mechanism had a handle that could be turned faster or slower to vary the number of images taken every second. The individual frames making up the film measured 2 x 2.5cm. The celluloid film was housed in a box at the top of the Cinematograph, unreeling at a speed dictated by the rate at which the cameraman turned the handle.

The Lumière brothers had patented their invention in February 1895, and had already staged three private showings, the first of them that March before the Society for the Encouragement of National Industry. The other two demonstrations of what the patent called an 'apparatus designed to obtain and display chronophotographic prints' took place in the offices of a science magazine and at the Sorbonne university in Paris.

Melodramatics
A scene from an American silent movie made in 1915 (above). Silent-film actors tended to exaggerate emotions and expressions, much as actors do on stage.

Building on the past
The brothers were the beneficiaries of a long series of prior inventions that had helped to pave the way for the coming of cinema. In 1825 an English doctor named John Ayrton Paris invented the thaumatrope, a disc with a different picture on each side attached to two lengths of string; when these were pulled, the disc revolved, causing the two pictures seemingly to combine in a single image. Seven years later the Belgian Joseph Plateau devised the phenakistiscope, a disc rimmed with a series of images that seemed to merge into one

A REAL-LIFE TRAGEDY

On 11 January, 1908, cinema's rapidly burgeoning popularity contributed to a tragedy in Barnsley, Yorkshire, in which 16 children were crushed to death. They were among a crowd that had turned out for a Saturday 'moving-picture show' organised in the town's Public Hall. The gallery became so packed that an usher shouted for some children to come down to better seats in the main body of the hall. The result was a stampede down the stairs in which the victims were trampled underfoot. All 16 children were under 10 years of age.

Having passed behind the lens, the film was then taken up on a second spool in another tightly sealed box below. Projection worked in much the same way, with the film unreeling frame by frame in front of the shutter. With illumination from a lightbox, each image was projected through the lens and onto the screen. It was this ability of the Cinematograph to project images onto a screen that permitted many people to experience simultaneous shared emotions in a darkened projection hall. The cinema was born.

Palace of pleasure
Refurbished for the Paris World's Fair in 1900, the Gaumont-Palace cinema in the city's Montmartre district (above) was the most sumptuous film theatre in Europe at the time, with seats for an audience of 3,400 people.

Mobile moviedrome
The Leilich Cinematograph (top), created by a Zurich-based carnival operator, moved from city to city as a fairground attraction. These mobile movie theatres gradually gave way to permanent venues.

To ensure that the projected image was fixed and uniform, the brothers developed the system of perforations in the film stock still used to this day. The claws on two sprocket wheels engaged with the circular holes to pull the film down in a succession of almost imperceptible jerks at a fixed rhythm of 16 frames a second, the minimum rate required to give an illusion of continuous motion and to avoid a flicker effect.

A franchise network

The Cinematograph exhibitions were an immediate success. Soon showings were running from 10 in the morning to 1.30 at night watched by some 2,000 people a day. The queues that built up before the Grand Café were so long that attendants had to be hired to keep the customers in order.

Yet for all the acclaim, Louis Lumière came to dismiss the Cinematograph as an 'invention without a future', a fairground attraction whose curiosity value would soon wear off. Before that happened, the brothers set out to make as much profit as possible from the enthusiasm inspired by the Paris shows. Proving themselves to be astute businessmen, they created a franchise scheme by which they sold exclusive rights to exhibit their films in other cities in France and abroad. In return for handing over 50 per cent of the receipts, the agents got the right to organise showings when and where they wished and to fix their own admission charges. They also had the loan of a Cinematograph and the necessary accessories, as well as the services of a trained operator provided by the brothers. These operators gave added topicality to the showings by filming views of the cities they found themselves in and projecting them there the same evening.

THE FIRST NATURE DOCUMENTARIES

Film production companies were quick to see the appeal of nature documentaries. In 1909 France's Pathé Studios set up a scientific popularisation department that produced a series of short films about micro-organisms. Two years later Éclair Studios launched a series called *Scientia*, shot in its premises in the Paris suburbs. The movements of creatures including alligators, sea anemones, spiders and chameleons, as well as the axolotl seen below, lent a strange silent poetry to these early productions.

MARRYING SOUND TO SIGHT

In 1908 the French composer Camille Saint-Saëns wrote the first film score, for a production entitled *The Assassination of the Duc de Guise* based on a play by Henri Lavedan, a member of the prestigious Académie Française. Featuring strings and some wind instruments, as well as a piano and harmonium, the composition was designed for an ensemble small enough to fit into a cramped orchestra pit while the film was being projected overhead. Eighteen minutes long, it was an immediate success; critics agreed that it perfectly complemented the action on screen.

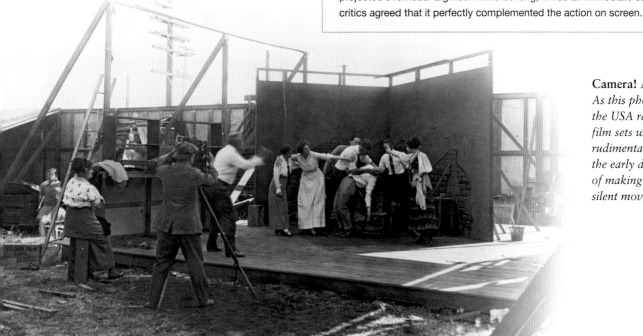

Camera! Action! *As this photo from the USA reveals, film sets were rudimentary in the early days of making silent movies.*

BIRTH OF THE NEWSREEL

In 1909 Charles Pathé produced the first cinema newsreel, the Pathé Journal, directed by Armand Verhyle and Lucien Doublon. The rival Gaumont film company was quick to respond, launching its own weekly news compendium in 1910. The features, which included studio reconstructions alongside live footage of topical stories, were shown either as show openers before the main film or else at the end of the evening's entertainment. They covered every conceivable type of subject matter, providing a significant new source of information for the general public.

First on film
A Pathé cameraman accompanied Bessie Coleman, the first Black American woman aviator, on a flight over Berlin.

Vision of the future
A scene from an early science-fiction film (below) shows one of the first attempts to imagine what spacesuits would look like.

To protect the exclusivity of the brothers' invention, the operators were bound by strict rules of secrecy and had orders never to let anyone else into the projection room. That could not prevent other enthusiasts who were not under contract to the Lumières from improvising devices of their own. Soon they were using them to film not just in France but also in many other parts of the world, from Australia and Austro-Hungary to Indochina and Mexico, where the cameras served to capture the lifestyles of distant peoples.

By late 1901 the Lumière brothers had a catalogue of film strips extending to 1,300 titles, but their business plan was proving too complicated to manage from their headquarters in Lyon. Instead, they started to sell their films and the machines to make and project them outright to their concessionaires.

Birth of an industry

The Cinematograph experience soon spread around Europe then the world. The 'Lumière performance' had its debut in Brussels on 29 February, 1896. On 1 April Emperor Franz Josef attended a showing in Vienna. The St Petersburg premiere took place in May, closely followed by New York in June and Shanghai and Mexico City in August.

Even though the brothers had the advantage of starting first, competitors using slightly different cameras and projection equipment soon made their presence felt. The pioneer film-maker Georges Méliès formed his own production company, Star Film, in 1896. Charles Pathé built vast studios at Vincennes, on the eastern fringes of Paris, and opened subsidiary branches around the world. Another big spender was Léon Gaumont, who started out in the Buttes-Chaumont district of northern Paris but by 1911 had 52 separate operations in France and abroad.

Britain was not slow to follow; by 1914 some 4,000 cinemas across the country were

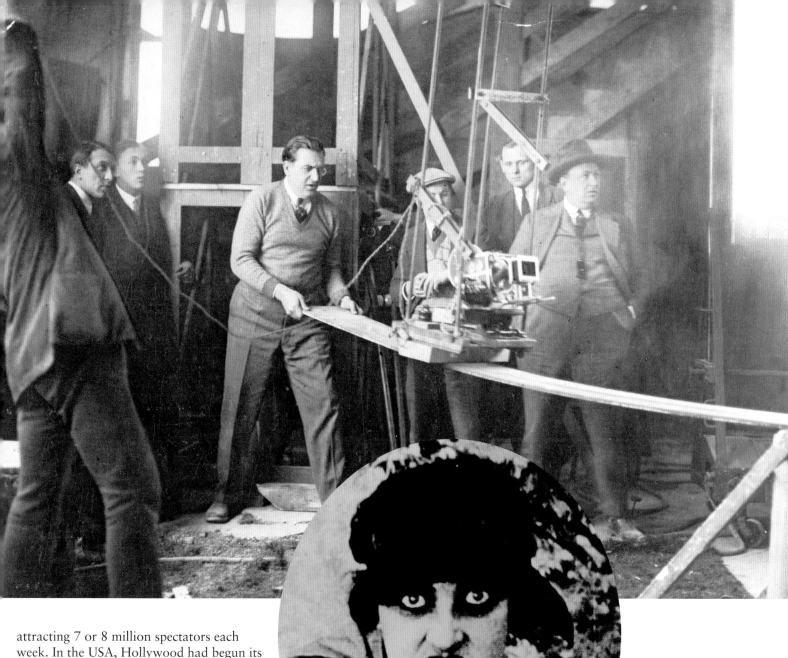

attracting 7 or 8 million spectators each week. In the USA, Hollywood had begun its rise to international fame. By the mid 1920s almost 50 million Americans were crowding the nation's 20,000 movie houses. The new attraction appealed to every social class, spreading out from fairgrounds to theatres, casinos and cafés. And even though the films themselves were silent, they were rarely shown without some musical accompaniment. Pianists or small orchestras were employed to heighten the emotional impact of the images. As early as 1897 the Lumière brothers were hiring a saxophone quartet to play in the course of their performances.

In the years that followed, permanent cinemas came to replace temporary venues, offering increasingly comfortable seating for film-goers. The first picture house in Britain was opened in Colne, Lancashire, in 1907 by a former magic-lantern showman; previously entrepreneurs with projectors had put on shows in hired rooms, touring the country in much the same manner as circus promoters. Britain had its own pioneer film-makers too,

notably London-based Robert W Paul and Birt Acres, who designed their own 35mm camera and made their first short film as early as 1895. In Blackburn, Lancashire, Sagar Mitchell and James Kenyon created an early production company that specialised in topical footage, newsreels and re-enactments. But despite developments in Britain and elsewhere, in the early days France was the true home of cinema: 90 per cent of all films shown around the world in the 1910s were made there.

Master at work
By the 1920s, when Fritz Lang (top) shot Metropolis, *the cinema was already becoming accepted as an art form. Many of the familiar genres of later years had already appeared by that time, including horror films. This image (inset) shows French star Musidora playing the lead role in Louis Feuillade's* Vampires, *filmed in 1915.*

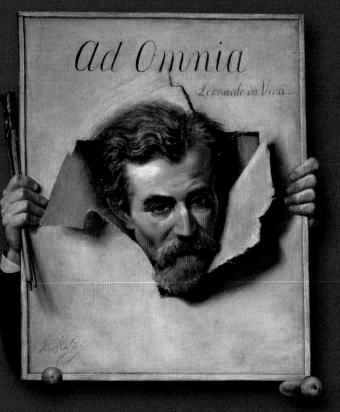

Magician of the silver screen

The Lumière brothers may have invented the Cinematograph and been the first to film real-life events, but it was Georges Méliès, the Jules Verne of the moving image, who can truly be said to have created cinema. He did so by turning his back on reality and instead conjuring up a world of fantasy. Thanks to the ingenuity of the technical effects Méliès employed, his work continues to fascinate viewers to this day.

Silent-cinema innovator
Unlike most pioneers of early cinema, who are remembered mainly for their technical advances, Georges Méliès was a creative talent whose innovations were mostly artistic. Over a 16-year period he made more than 500 films ranging in length from just one minute to about 40 minutes.

Méliès was born in Paris in December 1861, the son of a wealthy manufacturer of quality footwear. Like his brothers, he was initially expected to join the family business. After completing his studies and military service he travelled to London to learn English, but in practice he spent much of his time there visiting the city's music halls, where he was particularly taken with the performances of the stage magicians. Back in Paris in 1888, he used his inheritance to buy a theatre on the *grands boulevards* that specialised exclusively in conjuring acts.

Seven years later he sat spellbound through the first public screening of the Lumière Brothers' Cinematograph. When they refused to sell him their invention, Méliès instead bought the rights to an English projector that he converted so it worked also as a camera. He then used it to capture open-air 'views' – these early efforts were not yet known as 'films' – that he screened during matinee performances at his theatre.

Chance discoveries

Although Méliès began by recording the world around him, he was much more excited by special effects, which he stumbled upon by accident while shooting a Paris street scene. He later recorded the moment: 'One day while I was prosaically filming the Place de l'Opéra, the camera jammed, with unexpected results. It took a minute or so to unblock the film and get the machine going again, during which time the passers-by, buses and other traffic had, naturally enough, all moved about. When I projected the developed film, spliced together at the point where the jam had occurred, I found that the Madeleine–Bastille omnibus had been transformed into a hearse and male pedestrians had turned into women. I had discovered the stop-substitution effect, also known as the jump cut.' Méliès subsequently used the technique in *The Vanishing Lady* (1896), the very first film to use special effects, in which a woman turns into a skeleton. The film ends with the director himself appearing on camera to salute the audience.

From then on the pace quickened. That same year Méliès created what he called a 'Cinematograph film factory'– in modern parlance, a film production company. He was the sole proprietor of Star Film, whose slogan was 'The World within your Grasp'. In the following year he opened the world's first film studio in the garden of his house in the Parisian suburb of Montreuil-sur-Seine. Constructed to the same dimensions as his théatre – and boasting a stage complete with trap doors as well as storage for scenery and

---

FAKING THE CORONATION

In 1902 Méliès contrived to exhibit a film in Paris supposedly showing the coronation of Edward VII in Westminster Abbey on the day the ceremony took place. He did it by mocking up the event in advance in front of painted backcloths in his Paris studio, then employing a local laundryman of the same build and general appearance as Edward to stand in for the King. Far from taking offence, the British royal family subsequently invited the film-maker to visit Buckingham Palace.

other props – the studio was largely built of glass and was south-facing to catch the sun from 11 in the morning to 3 in the afternoon. A second studio had opened its doors before the year was out.

A virtual one-man show

Working almost non-stop, Méliès drew up scenarios for films, always leaving plenty of room for improvisation, as well as making all his own scenery and designing the costumes. He would be the first film-maker to produce hand-tinted movies, each frame coloured with his own brush. Besides directing the actors, he trained the camera operators, set up a film processing laboratory and explored an ever-more weird and wonderful variety of special effects. Like any good magician, he did not like to reveal his secrets, but the techniques he employed included dissolves, speeded-up and reversed motion, double exposures, vertical camera angles and freeze-frame. In *One-Man Band* he used multiple exposure to put seven versions of himself on screen at the same time. In *The Man with the Rubber Head* a bearded boffin uses a bellows to make his own severed head, set on a table, expand and contract like a balloon until it finally explodes. To make the head grow bigger and smaller,

Cinema magician
Méliès turns a sleeping woman into a butterfly in a special-effects sequence from one of his films (top). The scene above, of a condemned man losing his head on the guillotine, was shot for one of Méliès' early productions, but was cut from the final version in what must have been one of the earliest examples of screen censorship.

Méliès sat, with only his head showing, in a box secured to a trolley on rails facing a fixed camera. The closer the trolley came to the camera, the bigger the head appeared.

Reality and fantasy

Besides his mastery of fantasy and trick shots, Méliès also displayed a talent for re-creating headline news events. Using scale models, he mocked up the 1902 volcanic eruption of Mount Pelée in Martinique. He also filmed the explosion of the US battleship *Maine* – the event that triggered the Spanish-American War of 1898 – employing a miniature replica in a basin of water.

Méliès had his greatest hit in 1902 with *A Trip to the Moon*, inspired by a Jules Verne novel, which featured 30 scenes and 17 set changes, as well as an extensive cast of actors, among them ballerinas, cabaret singers and dancers from the Folies Bergère and other music halls. The film was a hit not just in France but also in the USA, where it was widely pirated and plagiarised. To protect his intellectual property, Méliès subsequently opened a Star Film branch office in New York. When the fairground showmen who were Méliès' principal clients complained about the

Trip to the Moon
Méliès made a 35-minute feature adapted from the Jules Verne novel From the Earth to the Moon. *The film charts the crazy adventures of Professor Barben-fouillis and his scheme to fire six men in a giant gun shell to the Moon. Méliès himself is shown (top left, inset) sketching the design for the film's best-known scene, when the projectile lands in the Moon's eye (above).*

cost and novelty of the project, the film-maker responded by himself staging showings at the Foire de Trône funfair in Paris. The venture drew big crowds, and soon fresh orders were flowing in.

A forgotten genius

By 1912 Méliès had produced some 520 films, but with the passage of time his work fell out of fashion. The rise of competitors like Pathé, Gaumont and Éclair Studios established realism as the dominant cinematic style, and Méliès never managed to adapt to the change in public tastes. Demand for his productions declined from 1910 on. By 1914 Star Film had gone out of business and he was bankrupt.

In 1923 Méliès was forced to sell his studios, and in a fit of despair he destroyed the

Fairy godmother
*Méliès' future wife Jehanne d'Alcy (right)
as she appeared as the fairy godmother
in* Cinderella *(1899). At six minutes'
playing time, this was the first of
Méliès' films to use up more than
100m of film stock.*

Early chorus line
*Méliès was one of the first
directors to feature a parade of
bathing beauties in his films,
as this scene from* A Trip to
the Moon *indicates. Mack
Sennett and other US
silent-cinema pioneers
would soon adopt
the same device.*

Rediscovered gems
*Toward the end of his life, Méliès – by
then largely forgotten by the public – was
delighted to be presented with some of
his lost films, freshly rediscovered by two
enthusiasts of early cinema (below).*

A PATIENT RECONSTRUCTION

Shortly before Georges Méliès' death, Henri Langlois – a passionate movie enthusiast who in 1936 would found the Paris Cinemathèque – managed to retrieve and restore a number of films by the pioneer film-maker. More than 200 of the 520 films made by Méliès between 1896 and 1912 have now been rescued. Almost every year new reels turn up in attics or cupboards, adding fresh material to the Star Film back catalogue.

negatives of many of his films. All but forgotten, he ended up running a toy stall at Montparnasse railway station in Paris, in partnership with Jehanne d'Alcy, once one of his actresses and by that time his second wife.

In 1929 copies of some of his films were rediscovered and the editor of a cinema revue organised a showing at the prestigious Salle Pleyel concert-hall in Paris. Méliès took the stage to receive the plaudits of the 2,500 people who attended the event. Two years later Louis Lumière pinned the Légion d'honneur on the breast of the old master, whose work would influence future generations of film-makers, George Lucas, Terry Gilliam and Tim Burton among them. Yet recognition came too late to revive his fortunes. With no further outlet for his talents, this 'great man who remained a little boy all his life', to quote French journalist Henri Jeanson, died poor and alone in a Paris rest home in 1938.

X-RAYS – 1895

Making the invisible visible

On 8 November, 1895, Wilhelm Röntgen stumbled on a mysterious, super-powerful new form of radiation. Not knowing the exact nature of the forces involved, he chose to give them the name of x-rays. Since that date Röntgen's rays have proved useful in many disciplines, providing science with an invaluable new tool.

Early radiometer
Invented by the British scientist William Crookes in 1875, this device (right) was designed to measure electromagnetic radiation. It consisted of a number of vanes attached to a central spindle, set inside an airtight glass bulb containing a partial vacuum. The vanes, coloured white on one side and black on the other, spun when exposed to light or infrared radiation, with the dark side retreating from the radiation source.

The exact circumstances of the discovery remain unclear, for Röntgen's research notes were burned after his death in 1923. What is known is that on that November day in 1895, while working as a physics professor at the University of Würzburg in Germany, he chanced upon a new form of radiation, more powerful than any yet known, and named it 'X' to indicate its unidentified nature. His discovery would not be properly understood for 20 years. The scientific world first learned of the rays on 22 December, when Röntgen published a paper entitled 'On a New Kind of Ray: A Preliminary Communication'. The discovery would win him the first Nobel prize for physics in 1901, an award that was thoroughly merited for the advances that the rays (which earlier researchers had observed but not identified) made possible in fields of science ranging from chemistry, microscopy, crystallography and astronomy to the medical disciplines of radiology and radiotherapy.

A chance discovery

Röntgen made his breakthrough while working on a different form of radiation: cathode rays, which would later prove to have important

long-term significance through their role in the invention of television. Cathode rays are in fact nothing more than electric currents (which is to say streams of electrons) passing not down a wire but instead through a vacuum created inside a glass tube (the vessels employed were known at the time as Crookes tubes). Emitted from a metallic electrode (the cathode), the electrons moved in a straight line for a few centimetres either to a second electrode that reabsorbed them (the anode) or to the glass wall of the tube.

In the event x-rays turned up purely by chance as an unintended side effect. Röntgen was projecting cathode rays onto a screen coated with a fluorescent substance, when he happened to notice that the screen was shimmering slightly even when the beam of electrons was not directed at it. Rather than simply trying to eliminate this unwanted phenomenon, he decided to investigate it further. First he put black cardboard between the tube and the screen, then a pile of books, but to no avail. The screen continued to shimmer, suggesting the presence of some unknown source of radiation, invisible to the naked eye, that was powerful enough to penetrate both obstacles.

Röntgen had discovered something, but he had no idea what it was. Seventeen years were to pass before the physicist Max von Laue

COMING CLOSE WITH CATHODE RAYS

In 1869 the co-inventor of the cathode-ray tube, German physicist Johann Wilhelm Hittorf, noticed that his photographic plates were sometimes flawed by shadows even when they had not been exposed to cathode rays, but he failed to follow up on the discovery. Nikola Tesla came close to discovering x-rays in 1894, when he observed that his tube emitted 'visible and invisible waves', which he wrongly interpreted as different types of cathode ray. Philipp Lenard, a student of Heinrich Hertz, was also not far off when he conducted experiments on the ability of cathode rays to penetrate different materials, but he did not specifically identify those that could penetrate as being an entirely new form of radiation.

identified the rays as powerful electromagnetic force-lines endowed with colossal energy equal to 100,000 times that of visible light.

A new way of imaging

Meanwhile, Röntgen continued to experiment. He substituted a photographic plate for the fluorescent screen and found that the rays left an impression upon it. Better still, he learned that they could capture the internal structure of objects interposed between the source of the x-rays and the plate, and that the clarity of the image depended on the density of the object, with metal providing a firmer outline than, for example, wood. He also noted that lead, which has a high atomic density, acted as a screen that blocked the rays.

It was while manipulating a piece of lead exposed to x-rays that he first noticed that his hand left an image on the photographic plate, blurred by the movement. More importantly, he saw that the flesh seemed transparent, while the denser bone matter showed up clearly. Realising the potential significance of this breakthrough he asked his wife, Anna Bertha, to pose for 20 minutes on 20 December, 1895, with her hand positioned between the tube and the photographic plate. By doing so he invented the x-ray photo, the first non-invasive form of medical imaging.

A dangerous new tool

Thereafter x-rays quickly spread around the world, for a time being embraced as something of a popular sensation and inspiring satirical poetry in *Punch* magazine. 'X-ray cabinets' appeared in towns and travelling fairs as a novel form of entertainment. The 1896 exhibition at London's Crystal Palace included an x-ray demonstration, as did the Exposition of the Electric Light Association in New York that same year, an event organised by Edison.

The medical establishment, too, proved enthusiastic. Within weeks of the publication of the image of Anna Röntgen's hand, a German dentist produced the first dental x-ray. In New Hampshire in February 1896 the American physician Dr Gilman Frost and his brother, physics Professor Edwin Frost, took the first medical x-rays of a boy's broken wrist. Also in the USA Thomas Edison invented the fluoroscope, a system combining an x-ray source and a fluorescent screen that made it possible to see the entire skeleton of a person set between the two. Similar techniques are in use to this day. But when Edison's assistant Clarence Dally died in 1903 of cancer linked to x-ray exposure, Edison immediately abandoned his researches into the technology.

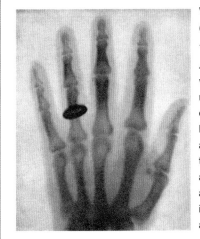

WHERE DO X-RAYS COME FROM?

X-rays are created by the flow of electrons in cathode rays when these are reabsorbed by matter in the form of the second electrode (the anode). It is now known that electrons emit x-rays as a result of the deceleration that occurs as they arrive at the anode, similar to the way in which a car that brakes hard dissipates its forward energy through friction and the heating of the tyres.

The first x-ray *Röntgen's pioneering 1895 x-ray of his wife's hand revealing the detail of the bone structure; the metal ring that she wore on her fourth finger is also clearly visible.*

Medical breakthrough *A coloured engraving from 1900 shows Röntgen examining an x-ray image of the lungs of a young patient suffering from pulmonary tuberculosis.*

CRYSTALLOGRAPHY

In 1912 the physicist Max von Laue demonstrated that x-rays are electromagnetic waves, just as visible light is. Shortly afterward, the father and son team of William Henry and William Lawrence Bragg showed that x-rays could be diffracted (turned aside) by the atomic structure of the materials they passed through. In doing so they laid the foundations of the science of crystallography, based on the fact that the diffraction of the rays by crystals – such as diamonds or salt – creates different patterns depending on the atomic structure of the crystal. The technique was gradually extended to organic substances such as haemoglobin, ultimately enabling scientists to penetrate the double-helix structure of DNA in 1953.

The whole skeleton
A full-body x-ray of a man reveals three lesions – one on each arm and another on the right knee.

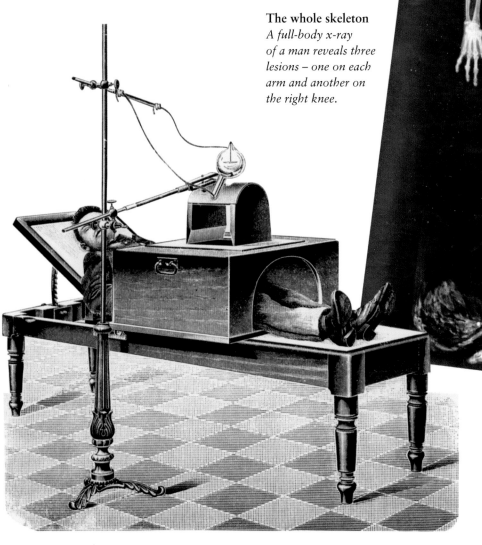

Early generator
A patient lies fully clothed in this illustration of an x-ray being taken in 1898.

'Don't talk to me about x-rays', he exclaimed, 'I'm afraid of them.' He had stumbled on the carcinogenic downside of the new technique, which over the decades that followed took a lethal toll of the scientists, doctors and photographers exposed to excessive radiation.

From the lab to the cosmos

In fact, Nikola Tesla, the renowned physicist and one-time employee of Edison, had signalled the risks associated with x-rays as early as 1897, noting in the course of animal experiments that they had damaging effects on living organisms. Gradually, scientists learned ways of protecting themselves and others from the dangers of radiation, even putting its lethal power to use against bacteria and other micro-organisms, as in the sterilisation of food or in radiotherapy treatment to destroy cancer tumours.

At the same time that the medical uses of x-rays were spreading around the world, physicists also started to use them, first to study the structure of solids, then to develop x-ray microscopes and new imaging techniques such as tomography and scanners. Eventually they even managed to track down the natural sources of x-rays – in the cosmos.

MACROCOSM AND MICROCOSM

X-rays are excellent tools for studying phenomena at both ends of the size spectrum. X-ray microscopes, first developed in the USA by Sterling Newberry in 1950, have enabled scientists to study matter on the almost unimaginably tiny scale of a millionth of a millimetre. At the other extreme the rays have permitted astronomers to gather information about stars, many of which eject streams of charged particles that emit x-rays in much the same way as in cathode tubes. This phenomenon can occur when galaxies collide, or through the absorption of matter by black holes, or when supernovas are created by giant stars exploding at the end of their cycle of evolution. To track these phenomena, NASA launched the Chandra satellite observatory in 1999 to scrutinise the cosmos for stellar x-ray sources.

Supernova

A composite image of supernova 1987A (above) made by combining separate photos, some of them taken by the Hubble Space Telescope, capturing visible light, and the others from the Chandra observatory, revealing x-rays. The combined image shows three rings of matter ejected by the star in its dying moments.

CT scanner

A patient about to enter a computed tomography scanner, a device that combines two-dimensional x-ray images to build up a three-dimensional picture.

The safety razor 1895

O ver the centuries men have used everything from sharks' teeth and seashells to sharpened bronze to shave off facial hair, but from the late 17th century such devices mostly gave way to 'cutthroat' razors, with an open steel blade that folded into the handle when not in use.

But as the nickname suggests, shaving remained a delicate business, so when the safety razor arrived in 1895, complete with replaceable blades, it found a ready market.

The newcomer was the brainchild of an American inventor, King Camp Gillette. It earned its 'safety' moniker from the metal shield that covered all but the very edge of the blade, reducing the risk of accidental cuts and nicks. Customers also liked the idea of throwing away used blades, the more so given that many feared that harmful corpuscles could pass from them into the bloodstream. Equipped with Gillette's invention, men no longer had to hone their razors each morning on a leather strap, or alternatively make a trip to the local barber to be shaved. This was a loss as well as a gain, since for every nervous customer who sat down in the barber's chair with images of Sweeney Todd in mind, there were others who would miss the social contact that came with the hot towels and soothing lotions, not to mention the kit of seven razors kept for the best clients – those who could afford to pay for a different blade for each day of the week.

The new safety razors found their natural home in the bathroom, increasingly a standard feature of houses from the 1880s on. There they lined up alongside shaving soap, bottles of eau de Cologne and shaving brushes, usually made from boars' bristles although badger hair was sometimes used. Men's toiletries thus took their place beside the powders and creams of the women of the household, who were still at the time barred the use of make-up – at least if they were concerned to maintain a reputation for respectability.

Despite being readily adopted, the first safety razors were in fact something of a disappointment, as the steel manufacturers of the day were unable to produce sharp enough blades. It was 1901 before a really satisfactory product reached the market, following the foundation of the Gillette Safety Company in Boston, USA.

Blade protection
A 1905 drawing shows the different parts that went to make up the safety razor. There was an obvious market for the invention, given that, then as now, most men shaved daily; the current figure for the USA is about 75 per cent. It has been calculated that the average man shaves some 330m² (almost 3,500sq ft) of skin surface over the course of a lifetime.

LATER DEVELOPMENTS

T he first safety razor specifically designed for women was the Gillette Milady, which appeared in 1915. Electric razors were developed by an American colonel, Jacob Schick, in 1928. Cordless battery-powered razors were introduced in the USA by the Remington Corporation in 1960. Eleven years later Gillette brought in double-bladed lightweight plastic razors, and three years after that, in 1974, Bic started producing disposable versions. Gillette responded the following year with the double-bladed disposable razor, and the brand remains a world leader to this day.

Speaking clock 1894

A clockmaker in Geneva named Casimir Sivan produced the ancestor of the speaking clock in 1894. The device he invented contained a phonograph cylinder with a recorded human voice that was only audible from close up.

The first speaking-clock telephone announcements were introduced in the USA in 1927. Britain got its service in 1936, after which residents of six major cities could dial TIM (the calling code was different for people on other exchanges) to hear a female telephonist informing them that, at the third stroke, the time would be the given number of hours, minutes and seconds 'precisely'. Until 1963 the voice was that of Ethel Jane Cain, who had won the grand sum of 10 guineas (£10.50p) in a competition to find 'the girl with the golden voice'. Initially the recording involved an entire roomful of motors, valves and glass discs, but in 1963 more modern equipment was substituted that captured the sound on a magnetic drum. The current digital system dates from 1984.

Fare counter
Early taxi meters like this one (above) would have been fixed to the outside of the vehicle, above the offside (driver's side) front wheel.

Taxi meters 1897

A taxi meter counts the number of times that the vehicle's wheels turn by measuring the rotation of the driveshaft connected to the gearbox. To arrive at the correct tariff it also records the time that elapses during the ride, thus taking into account the moments when the vehicle is not moving.

The device was invented in 1891 by a German, Wilhelm Bruhn. They were first put to use six years later, when a meter was fitted onto a Benz Victoria delivered to Friedrich Greiner, a haulage contractor from Stuttgart. Two years later Greiner purchased seven further cars equipped with meters, creating the world's first known fleet of motor taxis. In the years that followed petrol-powered cabs became familiar sights in almost all of the world's cities.

Rear-view mirrors 1896

As the number of cars on the roads increased, drivers felt a growing need to find out if they were being followed by vehicles travelling faster than they were, yet initially they could only do so by turning round to look over their shoulder. In 1896 an English army surgeon named John William Cockerill had the idea of fixing a small mirror above the windscreen. Alfred Faucher perfected the design in 1906. Adjustable mirrors were introduced by Pierre Stehle in 1949. External mirrors would follow, as would concave designs shaped to give a panoramic view (right).

THE START OF MASS TOURISM
Travel at the turn of the 20th century

In the late 19th century, before the coming of private cars, trains and steamships offered travellers the prospect of voyaging around the world at previously undreamed-of speeds. The more people sought to explore distant parts, the greater became their desire for creature comforts, fuelling a demand for sleeping cars and luxury liners.

Stepping out
A cartoon of 1857 (right) by the satirical artist Honoré Daumier shows Parisian tourists taken aback by the cold as they go for a paddle in the sea.

Well-dressed tourists
In the late 19th century travelling was not an excuse for casual clothes. These smartly dressed ladies (left) are emerging from an ice cave in the Bossons glacier near Chamonix in the Alps in 1865. Even at the turn of the 20th century, it was still de rigeur to look one's best on holiday. The group chatting on the boardwalk (top) are enjoying the sea air in Normandy in 1912.

The 19th century saw major technological advances in the field of transport that also widened people's horizons and broadened their view of the world. Writers set the pattern, flocking to the Mediterranean or seeking the enticements of the mystic East. Flaubert found his way to Egypt, James Joyce to Trieste, while Robert Louis Stevenson went further still, to the islands of the Pacific Ocean.

The Grand Tour and after

The first European tourists had been British aristocrats who, in the 18th century, set off on the Grand Tour of the Continent's cultural hot spots; the word 'tourism' would be coined from their example. The charms of the natural world, once considered a hostile environment, were soon recognised. It became fashionable to take trips to spas at Bath, Vichy or Baden-Baden; in time mountains, too, became poles of attraction, drawing visitors to Alpine sites such as Chamonix, with its Mer de Glace ('Sea of Ice') glacier, and the Great St Bernard Pass. Early visitors made the journey in carriages or on horseback, suffering mightily in the process. So when steamship and railway companies started offering their services for tours in the second half of the 19th century, they were welcomed with open arms.

Travelling for pleasure

In the very early days of railways trains were used mainly to move goods, but before long a new clientele of tourists took to the rails, attracted by evocative posters advertising foreign parts. Sea bathing became fashionable, encouraged by the notion that it had beneficial health effects, as had been pointed out a century earlier by the English physician Richard Russell. Following the opening of the railway from London in 1841, Brighton became Britain's most popular seaside resort, sporting a Royal Pavilion built by George IV and a theatre. Between 1841 and 1871 the town's population more than doubled to 90,011, making it the country's fastest growing town. The famous Palace Pier opened in 1891.

At the same time the first rack-and-pinion railways were opening up mountain resorts, such as St Gall in Switzerland, where the first winter-sports club came into being in 1871. In France the PLM (Paris-Lyon-Mediterranean) railway opened in 1857, providing a quick route to the Riviera. The enthusiasm for exotic destinations helped to create a market for

Upwardly mobile
Passengers ride a funicular up Vesuvius near Naples in the late 19th century. Nice and Venice were also popular destinations for the well-heeled traveller.

TRAVEL TRADE PIONEER

Derbyshire-born Thomas Cook organised the first excursion to be advertised publicly when, in 1841, he arranged for 570 temperance enthusiasts to travel by rail from Leicester to a rally in Loughborough, 11 miles away. The success of the venture persuaded him to try other initiatives; in 1851 he dispatched some 165,000 people to the Great Exhibition in London. By the 1860s Thomas Cook & Son were sending tourists as far afield as Switzerland, Egypt and the USA. In the 1870s they introduced so-called 'circular notes', which would later become better known as traveller's cheques (right).

FOLLOW THE GUIDE

In 1839 a German publisher named Karl Baedeker brought out a book entitled *Voyage along the Rhine from Mainz to Cologne: A Manual for Travellers in a Hurry*. The idea of guidebooks was not new; Baedeker got his inspiration from Byron's publisher, John Murray. But Baedeker's book proved so successful that it started a genre. A string of other Baedeker titles followed, winning a reputation for the accuracy of their information and the practical advice offered on transport and accommodation. English and French translations followed, and in time the very name 'Baedeker' became internationally synonymous with the guidebook genre.

Out on deck
A hand-tinted and tweaked photograph (above) shows passengers taking the air on the Empress Maria Theresa in 1895. The German liner was eventually sold to the Imperial Russian Navy; renamed the Ural, *it was sunk in the course of the Russo-Japanese War of 1904-5.*

Singing in the train

At the time local trains, whether in Britain, Europe or the USA, left much to be desired. They were dusty, windows rattled and buffet cars served food with little to recommend it. For long-distance travel, the situation improved dramatically after 1863, when the US inventor George Pullman introduced the Pioneer sleeping car, complete with convertible bunk-beds. Six years later – after many disputes between towns who all wanted the new line to pass their way – the first transcontinental railroad opened, linking the east and west coasts of the USA. In the years that followed the 'iron horse' opened up the West, turning distant California into a boom state, first economically and then for tourism.

Europe, too, was taking to the train. A Belgian entrepreneur named Georges Nagelmackers took up the concept of sleeping cars, creating the Compagnie Internationale des Wagons-Lits (CIWL) in 1876. Seven years later the firm introduced a luxury service consisting entirely of sleeping-cars, all decorated in mahogany and lacquer. This was the Orient Express, which linked Paris to Istanbul in just 80 hours and opened the way for the rolling palaces of the railways' golden age. Encouraged by success, the CIWL began investing in non-moving accommodation. Soon the era of first-class hotel chains had dawned, with the Grand opening in Rome in 1893, the Ritz in Paris in 1898 and London's Carlton Hotel the following year.

Bedding down
A young passenger is settled down for the night on board a Pullman sleeping car on America's Union Pacific Railroad in about 1870.

seasonal tourism, with the Côte d'Azur, Greece and Egypt all proving popular destinations as refuges from the northern winter. City breaks in Paris, Vienna or Rome were attractive spring and autumn choices, while the summer months drew visitors to spas and the Swiss mountain resorts, as well as to the seaside.

electricity and running water, making it akin to the most luxurious hotel. A new, faster generation of liners, with names like the *Etruria*, the *Umbria* and the *City of Paris*, entered service from the 1880s on. Soon they were attracting a clientele of aristocrats, celebrities and millionaires. Wealthy Americans took advantage of the vessels to immerse themselves in the culture of 'Old Europe'. Huge floating cities made the passage from Europe to New England in less than eight days, while steamships plied the seaways to North Africa, the Middle East, Asia, Australia and beyond, shrinking distances and swallowing frontiers. Even when air travel challenged their supremacy in the 1930s, the big ships continued to attract millions of passengers, as they still do today, riding the waves with elegance and pride.

Travelling in style
May McAvoy, a star of the silent screen, waits with her faithful companion to embark on a cruise liner in the 1920s. Shipping labels like those adorning her trunk were the mark of a well-travelled person.

In 1885 the Canadian Pacific Railway was inaugurated, giving Canada is own trans-continental railway. In 1891, in the reign of Tsar Alexander III, work got underway on the titanic public-works programme that created the Trans-Siberian Railway. Completed in 1916, the line stretched some 5,800 miles (9,200km) connecting Moscow to Vladivostok on Russia's Pacific coast, making it the world's longest railway until later extensions to the line took it even farther.

The great liners

The first wave of globalisation followed the spread of colonisation, increasing the demand for international travel. As steamships gradually replaced sailing boats, shipping lines providing regular, rapid connections sprang up to carry cargo and passengers alike. The first British transatlantic service was launched in 1838 and in 1869, with the opening of the Suez Canal, a postal service linking Europe and Asia became available.

In 1870, the American White Star Line's ship the *Oceanic* set a new standard for ocean travel. Its first-class cabins were set amidships, with the added amenities of large portholes,

THE JOYS OF CRUISING

Passengers on cruise ships prepared carefully for the voyage. They packed vast wardrobes, reckoning on up to four changes of clothes a day, including the obligatory evening dress. Arriving at the quayside with piles of luggage, they were met by a small army of porters waiting to carry the bags to the cabins. Champagne receptions greeted first-class guests to celebrate the ship's departure. On-board life was stylish and hectic, but for all the flowers, wine, caviar and general well-being it was hard to forget entirely the perils of long-distance sea travel. The *Titanic* was by no means the first luxury liner to go to a watery grave; its predecessors included the *Ville-du-Havre*, which went down in 1873 with the loss of 226 lives, and the *Bourgogne*, sunk in 1898 with 549 fatalities. Yet even disasters on that scale could not put off potential customers, who simply took the precaution of avoiding the lines thought to have the worst safety record.

Discovering the particles of which atoms are made

In 1896 Joseph John Thomson made a remarkable discovery: atoms could emit sub-particles of matter. The identification of electrons was the start of a process that by the end of the century would kill off the view, held since Classical times, that atoms were indivisible. It was nothing less than a revolution in the world of physics.

Test tube
Thomson made his breakthrough 1896 discovery in the course of experiments to study the behaviour of cathode rays in the presence of magnets and electrical fields. This glass tube (below) was a piece of the apparatus used in his experiments.

The discovery of electrons marked a turning point in science. Once physicists realised that they were components of atoms, the myth of the indivisible atom, passed down from the days of ancient Greece and accepted by all previous scientists from Lavoisier to Dalton, was exploded once and for all. The vehicles for the breakthrough were cathode rays, which a year before had already revealed the existence of x-rays. When physicists first began to investigate cathode rays, they naturally sought to know what they consisted of: waves, atoms and molecules were all considered and rejected. The question became urgent when Röntgen went on to demonstrate that they could in their turn produce ultra-powerful x-rays, suggesting that scientists were confronting a chain of related phenomena whose logic they did not comprehend.

Making the breakthrough
Manchester-born Joseph John Thomson put the process in motion in 1896 when he undertook three crucial experiments on cathode rays. From the first two experiments he learned that cathode rays were composed of a single type of particle, carrying a negative charge. The third enabled him to demonstrate that the particles had mass, the fundamental characteristic of matter. The conclusion he drew was that, unlike light or x-rays, cathode rays were made up of particles of matter. Thomson even managed to estimate the size of the mass involved, reckoning that it was at least a thousand times less than that of the smallest known atom, hydrogen. But he also realised that the particles could not simply be a new type of atom, because they shared none of the requisite characteristics: specifically, they could not be chemically isolated, nor could they be made to react with other atoms.

Subatomic particles
Thomson noted, in an article published in 1897, that no matter how the cathode rays were produced, all the particles of which they were composed had the same mass. This was independent not only of the speed at which the rays were travelling but also of the type of

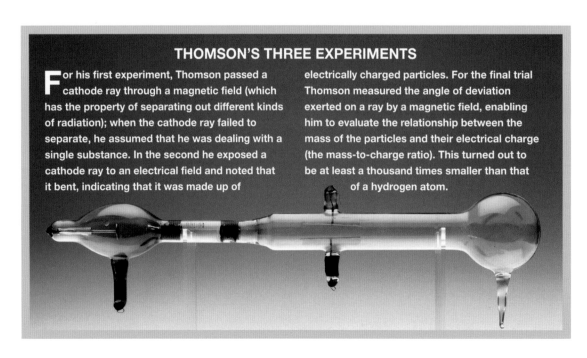

THOMSON'S THREE EXPERIMENTS

For his first experiment, Thomson passed a cathode ray through a magnetic field (which has the property of separating out different kinds of radiation); when the cathode ray failed to separate, he assumed that he was dealing with a single substance. In the second he exposed a cathode ray to an electrical field and noted that it bent, indicating that it was made up of electrically charged particles. For the final trial Thomson measured the angle of deviation exerted on a ray by a magnetic field, enabling him to evaluate the relationship between the mass of the particles and their electrical charge (the mass-to-charge ratio). This turned out to be at least a thousand times smaller than that of a hydrogen atom.

electrode used. In other words, the newly discovered particles were all the same whatever type of matter (in the form of the electrodes) ejected them. The obvious conclusion was that cathode rays must consist of particles of matter that, being too light in weight and too atypical to be atoms themselves, could only be particles of atoms. So atoms must be divisible. The idea came as a bolt from the blue. Thomson initially referred to the particles as 'corpuscles', but the term 'electron' was soon substituted. It was coined in 1894 by the Irish physicist George Stoney, who had made surprisingly accurate predictions about the existence of 'atoms of electricity' within matter, an idea that no doubt helped to inspire Thomson.

Scientific luminaries
Joseph John Thomson and Ernest Rutherford, both Nobel prize-winners, engaged in conversation on the eve of the Second World War.

FINDING A NAME

The word 'electron' was coined by George Stoney to describe what he called 'this most remarkable fundamental unit of electricity'. The –*on* ending was applied to other subatomic particles as they were discovered, among them the photon and neutron.

Rutherford enters the stage

As it turned out, physics had other surprises in store in those closing years of the 19th century. The sequel to the deconstruction of the Classical Greek concept of the atom came from the other major discovery associated with cathode rays, namely x-rays. In the same year in which Thomson identified electrons, the French scientist Henri Becquerel chanced upon the radioactivity of uranium while seeking to explain the origins of x-rays. This was the context in which, in 1899, a young physicist from New Zealand named Ernest Rutherford, later to become one of the giants of science, found a new flaw in the supposed impenetrability of atoms. He discovered that radioactive materials emitted not just x-rays but also particles of matter. Bizarrely, it seemed that radioactive material that was theoretically composed of a single type of atom was in fact emitting particles of an entirely different type of matter. The problem was a real brain-teaser.

Downward spiral
An image made in a bubble chamber at the Lawrence Berkeley Laboratory in California, following the path of an electron (below). As the electron ionises the liquid hydrogen contained within the chamber, it loses energy and the diameter of its orbit diminishes, creating the spiral effect.

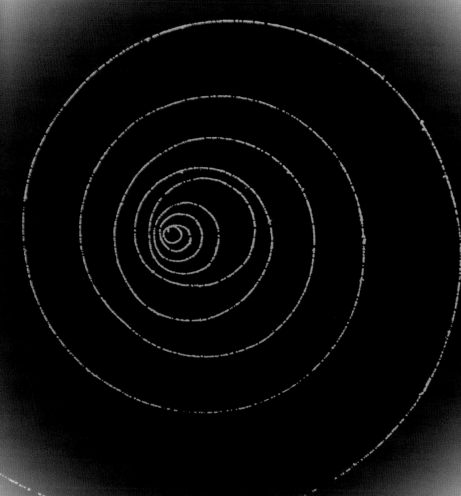

Beta particles

Moving very rapidly, beta particles travel in a perfectly straight line across a bubble chamber (right).

Significant traces

Alpha particles emitted by atoms of different masses leave distinctive traces in a bubble chamber (below). As the one shown in yellow releases a proton that moves off to the right (red line), it causes the particle itself to veer a little off its trajectory.

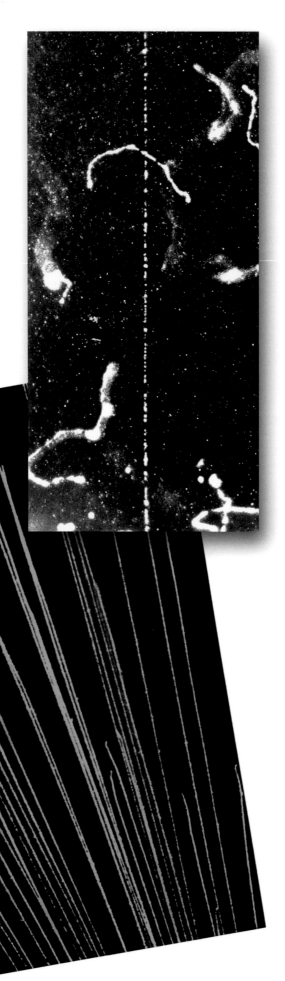

Alpha and beta particles

Rutherford had stumbled on the paradox while testing the penetrative power of the radiation emitted by uranium. He placed a sample of pure uranium against fine strips of metal of varying thicknesses, and soon came to realise that the forces he was dealing with were not x-rays alone. While the latter could pass clean through the obstacles, two other types of radiation that he detected could not, being effectively blocked by the strips. He named them alpha and beta; the beta particles could penetrate a little further than the alpha.

At that point Rutherford was still unable to define the nature of the two new types of radiation, but he gradually developed a hypothesis that they involved different types of particle. The alpha subatomic particles, which had the least penetration, were bigger and therefore heavier, while the beta particles were smaller and lighter.

Towards a model of the atom

Part of Rutherford's hypothesis was confirmed a year later when Henri Becquerel discovered that beta particles behaved in the same way as the electrons in cathode rays and also had the same mass. He was thus able to kill two birds with one stone by demonstrating that beta particles and electrons were in fact one and the same thing. On the one hand this confirmed Thomson's proposal that electrons came from atoms, since atoms of pure uranium emitted them. On the other, by demonstrating that beta radiation was composed of particles of matter, Becquerel lent support to Rutherford's view that the same applied to alpha radiation.

Questions still remained as to what form of matter was involved. Working with another future Nobel prize-winner, Frederick Soddy, Rutherford found an answer in 1902: alpha particles were helium atoms, implying that

---

GAMMA RAYS

In 1900 Paul Villard, a French physicist working with radium, discovered the third and last type of radiation emitted by radioactive material: gamma rays. He established that they were waves, like x-rays, rather than a flow of particles, like alpha and beta rays, from the fact that they were unaffected by magnetic and electrical fields. He also discovered that they had much greater powers of penetration than x-rays: in fact they represent the most energetic form of electromagnetic radiation.

THE ATOM

Atoms could be described as resembling a planetary system, with a cloud of tiny electrons orbiting a central nucleus formed of two types of particles, protons and neutrons. These last, whose number determines the nature of the atom, are held together by 'nuclear' forces, while the electrons are bound to the nucleus by electro-magnetic forces. The latter act as hooks, enabling atoms to link up to form molecules.

Patterns of life
Above: A computer-generated image of the orbital path of electrons in a hydrogen atom, showing their probable positioning.
Right: The electromagnetic disturbance caused by the entry (from the left) of a neutrino into a liquid-hydrogen bubble chamber.
Top: Nils Bohr's celebrated model of the atom.

uranium atoms somehow transmuted into helium and were then ejected. From this discovery Rutherford went on in 1909 to elaborate the first working model of the internal structure of atoms, refined four years later by Nils Bohr. The work marked the birth of quantum physics, one of the great revolutionary forces of modern science that in time would transform not just physics but also chemistry, astronomy and medicine. It would even impact on daily life thanks to one of its best-known applications – the laser.

Sport crosses frontiers

To bring the peoples of the world together in peace, united by shared sporting values, was the goal of Baron Pierre de Coubertin when he relaunched the Olympic Games after a gap of some 1,500 years. His global perspective contrasted strongly with the heightened nationalism typical of the late 19th century.

Pierre de Frédy, Baron de Coubertin, was a French educationalist with deeply held views about the important role that sport could play in the physical, intellectual, artistic and civic development of young people all over the world. In 1894 he gathered together representatives of a dozen or so countries in Paris and proposed reviving the Olympic Games. The delegates of the International Olympic Committee, founded that year, chose Athens as the seat of the first concourse of modern times. The Athens Games were restricted to male amateurs; women first competed in the 1900 event, held in Paris.

One important reason for the success of the Athens Games was the renovation of the Panathenaic Stadium, a structure dating back to Classical times situated in the heart of the city. It was refurbished in marble, and its seating capacity was raised to 69,000. The opening and closing ceremonies were staged there, along with some athletic events. Other sports were housed in venues nearby.

Sporting legends

Eighty-one athletes from a dozen countries responded to the invitation, competing against 230 local competitors in 40 events representing nine different fields of sport.

Olympic champions
The victorious German gymnastic team wearing crowns of laurel (above), having won both team events and three individual titles.

On your marks
Competitors adopt a variety of starting positions for the 100m sprint at the 1896 Athens Olympics (below). Longines stop-watches (right), accurate to roughly a fifth of a second, were used to time the race, which was won by the American Tom Burke in 12.0 seconds.

Link to the past
Restored by the architect Ernst Ziller for the 1896 Olympic Games, the Panathenaic Stadium in Athens was originally constructed by Lycurgus in the 4th century BC and enlarged by Herodes Atticus in AD 140.

Some 40 accredited European journalists covered the occasion. Times for the running races were recorded with the aid of a chronograph provided by the Swiss clockmaker Longines. Track and field events provided the main focus, followed by gymnastics, swimming, cycling and tennis. Wrestling and weight-lifting were represented, but boxing was considered too brutal and was omitted. Professionals only competed in a special fencing event reserved for masters of arms. Coubertin insisted on the inclusion of shooting, a sport he practised himself.

The American James Connolly was the first individual in the modern era to take Olympic gold when he won the triple jump. But the star of the games was a Greek shepherd, Spiridon Louys, who triumphed in the marathon.

Reaching out to the world

The Olympic Games has steadily grown in size and scope. The Winter Olympics were held for the first time in 1924 at Chamonix in France. Until 1992 they were scheduled for leap years, like the summer games, but the next contest was arranged for 1994, so that the two events are now separated by a two-year interval.

London hosted its first games in 1908, stepping in for Rome, which pulled out as host following an eruption of Mount Vesuvius that required a huge relief effort. In 1948 London also hosted the first Olympics after the Second World War. Known as the 'austerity Games', they were nonetheless a great success.

Olympic hero
The black American athlete Jesse Owens (right) competing in the 1936 Olympics in Berlin. He won four gold medals, setting a world record of 10.3 seconds in the 100m final. Owens' success dealt a blow to Nazi theories of Aryan racial supremacy.

The impact of the Olympics expanded vastly with the advent of television, sometimes to the detriment of the values championed by the Olympic movement. Politics, which made its ugly presence felt in the Nazi-sponsored 1936 Berlin games, raised its head again in the terrorist attack at Munich in 1972 and in the boycotts led by the USA and USSR, in 1980 and 1984, negating the ancient ideal of the Olympic truce. In more recent years the growing importance of money, as seen in the erosion of amateurism, the growing role of sponsorship and the huge sums generated by television rights, have all inevitably had consequences for the spirit of the Games.

THE MARATHON

The race commemorates the feat of Pheidippides who, according to legend, died of exhaustion after bringing news to Athens of the Greek victory over invading Persian forces at the Battle of Marathon in 490 BC.

THE CLASSICAL HERITAGE

The ancient Greek Olympic Games were held in honour of Zeus, king of the gods. They were staged every four years from 776 BC on, and featured athletic contests, equestrian events and wrestling. Winners were crowned with wreaths of laurel and were greeted with widespread acclaim. The Roman Emperor Theodosius banned the games as relics of paganism in AD 394.

An earth-shaking discovery

In March 1896 the French physicist Henri Becquerel discovered that some substances naturally emit x-rays. A young couple, Pierre and Marie Curie, took up the task of investigating the phenomenon. Soon a scientific revolution was in the making.

At the time of his breakthrough, the 43-year-old Henri Becquerel was professor of physics at Paris's Museum of Natural History. The announcement of the discovery of natural radioactivity took place in two stages. Becquerel made a first brief statement to the Paris Academy of Sciences on 24 February, 1896, then provided a fuller description a week later. Between the two communications came a reassessment of the evidence that effectively upgraded its significance, signalling the difference between an interesting discovery and a scientific revolution.

The process had got under way that January when Becquerel, who had been working on phenomena associated with phosphorescence and fluorescence, turned his attention to x-rays, discovered by the German physicist Wilhelm Röntgen the previous year. Röntgen had shown that x-rays were produced in cathode ray tubes, and the discovery set the scientific world alight as researchers competed to explain their presence. Becquerel was one of the seekers, and his findings were to have a dramatic impact on the world of physics.

Radiation pioneer
Seen here in his laboratory in the 1890s (right), the French physicist Henri Becquerel contributed indirectly to the development of radiotherapy in the treatment of cancer as the first person to notice that radium, which he often carried about in his pocket, could cause burns.

Radiant glow
A colourised image of radioactive alpha particles emitted by radium (below).

A helping hand

Becquerel's friend Henri Poincaré, himself a physicist and mathematician of genius, helped to set him on the right track. Poincaré had come to believe that x-rays were somehow linked to the phenomenon of fluorescence, for Röntgen's experiment had shown that their production was accompanied by a fluorescent effect on the glass Crookes tubes used in the experiments. On 30 January, 1896, Poincaré published an article in which he posed the question: 'Might one not wonder whether all bodies whose fluorescence is sufficiently intense emit Röntgen's x-rays as well as rays of light?' Becquerel took up the challenge, seeking to answer Poincaré's question.

WHAT IS RADIOACTIVITY

Radioactivity is energy lost through the process of decay within the nuclei of certain atoms, by which they transmute into atoms of a different type. When atoms form, some have nuclei with an unstable structure, like a badly balanced pile of fruit or layers of different bands of snow; the least disturbance can be enough to cause an avalanche that transforms the unstable nucleus of, for example, uranium into more stable lead. This transmutation, which takes place step by step, is punctuated by the emission of neutrons and protons (in the case of alpha radiation), of electrons (beta radiation), or of high-energy photons (gamma radiation).

The first step was to try to detect emission of x-rays by fluorescent substances. Immediately, though, a problem emerged, for the rays could only be identified by the impression they left on a photographic plate kept in complete darkness, while fluorescent substances had to be exposed to light to demonstrate their fluorescence. To get around this, Becquerel devised an experiment in which the plates were hermetically sealed in black paper and placed under a strip containing the fluorescent substance that he wished to test, which was then exposed to sunlight. He reasoned that if x-rays were produced, they would traverse the black coating, leaving as evidence an image of the strip on the plate.

Twin discoveries

In the course of that February Becquerel experimented with a number of substances, among them uranium salts. They turned out to be the only ones that left a mark on the plates. He was convinced that he had found the first natural source of x-rays, but, following Poincaré, he initially thought that the reason they did so was their fluorescence. In the note that he read out to the Academy on 24 February he wrote: 'One must conclude from these experiments that the phosphorescent substance in question [the uranium salt] emits rays that pass through the opaque paper.'

Three days later, Becquerel realised that photographic plates laid under fluorescent strips that had not been exposed to the light were also clouded. This indicated that the phenomenon had nothing to do with fluorescence, which only manifested itself after exposure to light. He had discovered that uranium naturally emitted a form of radiation even in the absence of light. In short, he had stumbled on radioactivity.

Cue the Curies

The discovery raised important theoretical questions. How could a substance produce such activity apparently spontaneously? Where did the energy come from? A young married couple living in Paris, both of them scientists, flung themselves into the search for answers. Pierre and Marie Curie started out with the hypothesis that the radioactivity came from the uranium atoms, even though no-one at the time really understood the nature of those atoms nor why they should be capable of delivering energy. For want of any clear idea of where their research might lead, the Curies sought to study radioactivity in great detail, but to do so they needed a quantity of pure uranium sufficient to support their researches.

COME FROM THE COSMOS

The origin of radioactivity is cosmic in that radioactive elements, like almost all the matter making up our world, are formed in the stars, then dispersed through space when stars die. Meteorites that fall to earth from the Solar System's asteroid belt, or from neighbouring planets like Mars, can also contain such elements, leading astronomers to suggest that the entire system, including the Sun itself, might have drawn some of its matter from a dead star.

Remnant of a supernova
Supernovas occur when dying stars explode. In February 1987 the first in four centuries to be visible to the naked eye occurred in the Large Magellanic Cloud, enabling scientists to test the theory, first advanced in the 1960s, that the blasts emit radioactive material.

PHOSPHORESCENCE VERSUS FLUORESCENCE

Phosphorescent materials absorb light to which they have been exposed then release it gradually over a period of time. Fluorescent materials are quicker-acting, emitting light while still illuminated.

URANIUM PRODUCTION

The first mine to provide uranium-bearing minerals for the Curies was the Jachymov facility, now in the Czech Republic. Since the 1940s, with the development of atomic bombs and nuclear reactors, extraction of uranium ore has spread around the world. Annual global production is now between 35,000 and 45,000 tonnes, with the principal producers being Canada (25 per cent of world production), Australia (19 per cent) and Kazakhstan (13 per cent), followed by Niger, Russia and Namibia.

Marie Curie at work
The pioneering scientist photographed in her laboratory in an annex of the Sorbonne university in Paris in 1912 (right).

Uranium ore
Pitchblende (right), also known as uraninite, is the world's principal source of uranium. It was the raw material used by the Curies in their discovery of radium and polonium.

Measuring device
Invented by Pierre Curie and his brother Jacques, this ionisation chamber (below) incorporated an electrometer that used quartz crystals to measure electrical charges, enabling scientists to measure the radioactivity of substances placed within it.

Slaves to science

For years the pair slaved away in a laboratory at the School of Industrial Physics and Chemistry in Paris, separating out the element from uranium-bearing ores, principally pitchblende. Typically, it took 3 tonnes of pitchblende to yield 1g of pure uranium.

In late 1897 Marie began a series of experiments designed to measure the effects of radioactive uranium on the atmosphere. Using a device for measuring electric charges that Pierre and his brother Jacques had developed some years earlier, she discovered that radiation 'ionised' the air, whose molecules lost their usual resistant properties and instead began to conduct electricity. The finding suggested to her that the atoms present in the molecules lost electrons when they were struck by the rays. Next she tried testing other materials and found that thorium also emitted similar rays, indicating that radioactivity – a term Marie coined at this time – was a distinct natural phenomenon and was not limited exclusively to uranium.

Encouraged by these findings, the Curies redoubled their efforts and soon chanced on another phenomenon. They discovered that the residue left behind by the purification process they applied to the pitchblende and chalcolite used in their experiments was actually more radioactive than the uranium itself. They concluded that there must be other radioactive elements within it, which they duly set about trying to isolate. By July 1898 they had enough evidence to be able to announce the discovery of polonium, named in honour of Marie's native Poland. That December, they announced their discovery of radium.

Undergoing radiotherapy

Radiotherapy treatment uses ionising radiation to counter cancer (below). More than half of all cancer patients now undergo radiotherapy, either on its own or in combination with chemotherapy and/or surgical intervention.

A triple Nobel

In 1903 the Curies and Henri Becquerel were jointly awarded the Nobel prize for physics. Their discoveries opened the door to atomic and nuclear physics, which in turn led to the discovery of the elementary particles, the building-blocks of matter, as well as of artificial radioactivity, first described in 1934 by their daughter Irène Curie and her husband Frédéric Joliot. In future years radioactivity would provide a treatment for cancer (in the form of radiotherapy) and techniques for exploring the human body (nuclear medicine) as well as for dating fossilised organisms (carbon-14 dating). It would also provide an important source of energy in the form of nuclear power – and, of course, radioactive materials are at the heart of nuclear bombs.

Radium pioneer

Marie Curie was the first female scientist of genius to be recognised as such by her peers – although not without a struggle along the way. She was the first woman to become a university professor in France and she received two Nobel prizes – one for physics and one for chemistry – for her work on radioactivity. Her legacy to science is immense.

Pathfinder

A portrait of Marie Curie (above) which appeared in a book about radium published in 1913, when she was 45 years old. The image to the right shows an x-ray photograph of her purse taken in 1900, revealing a key and coin inside.

Marie Sklodowska was born in 1867 to a family of teachers in Warsaw. She had a troubled childhood, marked by privation and the death successively of her sister Zofia and then her mother. At a time when eastern Poland was under Russian control, she pursued her scientific studies at the clandestine Floating University – so called because students moved from place to place to avoid detection. At the age of 24 she left for France, where she studied physics, chemistry and mathematics at the Sorbonne in Paris. In 1894 she met a young teacher at the university's School of Industrial Chemistry and Physics by the name of Pierre Curie. She helped him with his researches, and soon the pair were in love. Marie Sklodowska became Marie Curie on 26 July, 1895.

A union of hearts and minds

In 1896, when Henri Becquerel first stumbled upon radioactivity (a term coined by Marie in 1898), the young woman decided to make it the subject of her thesis. Over the following years she devoted much of her time and energy to the project. Eventually her health was sacrificed, too, for in the long run the research proved noxious.

Over the next two years she achieved remarkable results that at first the scientific community – unused to the idea of a woman making major discoveries in her own right – insisted on attributing to her husband Pierre. In fact Pierre gave up his own researches to assist his wife, working with her in perfect intellectual harmony. In 1897 their first child,

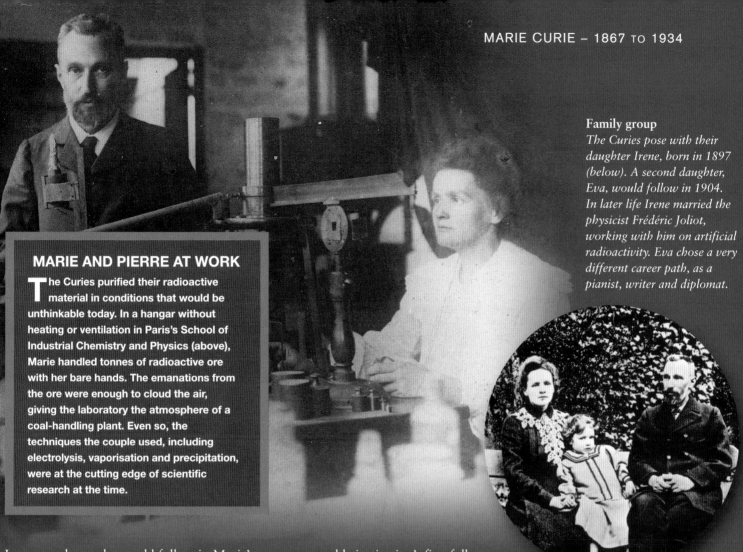

Family group
The Curies pose with their daughter Irene, born in 1897 (below). A second daughter, Eva, would follow in 1904. In later life Irene married the physicist Frédéric Joliot, working with him on artificial radioactivity. Eva chose a very different career path, as a pianist, writer and diplomat.

MARIE AND PIERRE AT WORK

The Curies purified their radioactive material in conditions that would be unthinkable today. In a hangar without heating or ventilation in Paris's School of Industrial Chemistry and Physics (above), Marie handled tonnes of radioactive ore with her bare hands. The emanations from the ore were enough to cloud the air, giving the laboratory the atmosphere of a coal-handling plant. Even so, the techniques the couple used, including electrolysis, vaporisation and precipitation, were at the cutting edge of scientific research at the time.

Irene, was born; she would follow in Marie's footsteps, becoming a Nobel prize-winner herself in 1935 when she shared the award for physics with her husband Frédéric Joliot for their joint discovery of artificial radioactivity.

The pain of bereavement

Between April and December 1898, the Curies presented the results of Marie's researches to the French Academy of Sciences, describing the properties of radioactivity and culminating with news of the discovery of polonium and radium. In 1903 Marie was awarded a doctorate by the Sorbonne; in the same year, she and Pierre won the Nobel prize for physics, with Henri Becquerel. It was the first time that a Nobel prize had been awarded to a woman. In the wake of the honour Pierre was given a professorship at the Sorbonne and Marie became director of research in his laboratory.

Everything seemed to be going well when, in 1906, tragedy struck. Crossing a Paris street in the rain, Pierre slipped and fell under a horse-drawn vehicle; he died instantly. Devastated by the loss, Marie wrote that she had been left 'incurably and desperately lonely'. She threw herself into her work with the force of despair. In an unprecedented move the Sorbonne offered her the chair of physics left vacant by Pierre, making her that venerable institution's first full woman professor. As the world's leading expert on radioactivity, she was awarded a second Nobel prize, this time for chemistry, in 1911.

Paying the price

Marie then launched the great project of her later years: the creation of an Institute of Radium, dedicated to research into radioactivity and its use in treating cancers, for by then Marie and Dr Émile Roux had between them discovered that the rays emitted by radioactive elements could kill cancer cells. The Institute opened its doors in 1914.

During the First World War Marie organised mobile x-ray units, known to the troops as 'little Curies', to help medical staff at the Front. In 1921 and 1929 she toured the USA, raising funds for the Institute. By then, though, the ill effects of her exposure to radiation were taking their toll. Having handled radioactive materials unprotected for decades, she finally succumbed to leukemia on 4 July, 1934, leaving her legend to live on after her.

Contributing to the war effort
Marie Curie at the wheel of a mobile x-ray vehicle during the First World War (below). When hostilities broke out in 1914, she championed the use of mobile radiography units to treat wounded soldiers.

The chemistry of the life force

For centuries people had speculated about the mysterious life force that enabled living organisms to breathe and to nourish themselves, drawing sustenance from the air, water and minerals. It was a German chemist, Eduard Buchner, who finally cast light on the important role that enzymes play in the process.

Baby food
Seen through an electron microscope, crystals of lactose and lactic acid form fronds (left). Babies secrete the enzyme lactase in their intestines, which enables them to separate and digest the fronds. As babies grow, their production of lactase diminishes, leading them to need other food in their diet.

Vitalism – the belief that living matter could not be explained by the laws of chemistry alone – was still widespread in 19th-century medical circles. Proponents of the theory insisted that a life force was necessary to animate inanimate matter, and that this force lay at the root of complex processes like digestion and fermentation. The life force idea came to be challenged in the 19th century by a series of discoveries that culminated in an understanding of enzymes and the important role that they play.

Enzymes, which Louis Pasteur initially called 'ferments', are natural catalysts that facilitate and accelerate chemical reactions at relatively low temperatures without themselves being affected by those reactions. The first to be discovered was diastase, isolated in barley germinated by the French chemist Anselme Payen in 1833, which broke down large starch molecules. Shortly afterwards the Swede Jöns Jacob Berzelius formulated the hypothesis that chemical reactions were more common than had previously been thought in living organisms.

A cascade of discoveries

Research carried out in the years that followed seemed to bear Berzelius out. A variety of different biological catalysts were discovered to be at work in living tissue. In 1836 the German physiologist Theodor Schwann isolated pepsin, which digests proteins, in a pig's stomach. The next year his compatriots Friedrich Wöhler and Justus Liebig extracted emulsine, capable of emulsifying oil, from bitter almonds. In 1840 Claude Bernard tracked down lipase, which breaks down lipids, in the pancreas. Twenty years later it was the turn of Marcellin Berthelot, who found

---

A FIRST STEP

The French scientist René Antoine Ferchault de Réaumur provided the first experimental evidence of the activity of enzymes in the 18th century, when he demonstrated the role that gastric juices play in the digestion of meat by birds of prey.

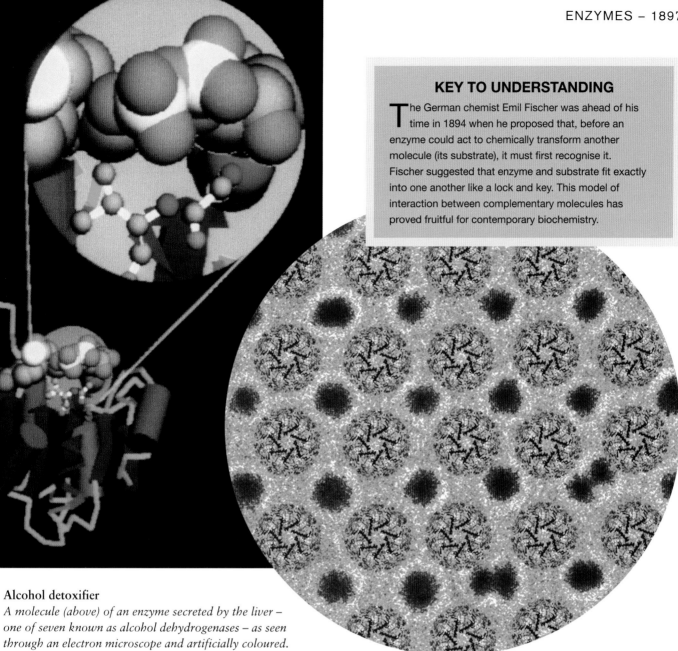

Alcohol detoxifier
A molecule (above) of an enzyme secreted by the liver – one of seven known as alcohol dehydrogenases – as seen through an electron microscope and artificially coloured.

invertase in yeast, where it played an active part in the hydrolysis of sucrose (table sugar).

In 1876 the German scientist Wilhelm Kühne discovered trypsin, which attacks proteins, in the pancreatic juices. He coined the word 'enzyme' (from the Greek *en*, 'in', and *zume*, 'leaven') as an alternative to 'ferment' or 'diastase', which until then had also sometimes been used generically to describe the whole class of substances. Each of the enzymes discovered to that point seemed specific to a single chemical reaction, but it was still unclear exactly how they worked. Above all there was disagreement about their role in the vital processes. The subject of alcoholic fermentation was contested especially fiercely. Claude Bernard suggested that a soluble fermentation agent secreted by yeast – in other words, an enzyme – could be sufficient to set in motion the breakdown of vital elements into

assimilable parts. But Louis Pasteur clung to the vitalist hypothesis, insisting on the fundamental role played by yeast's living cells.

A potent extract

In October 1896 the German chemist Eduard Buchner visited his brother Hans, a Munich-based bacteriologist. For some years the two men had shared an interest in the biological activity of yeasts. Their goal was to pulverize them in the hope of extracting a cell-free 'press juice'. To do so Eduard devised a new procedure, combining the cells with sand and powdered silica, covering the paste in a cloth and then using a hydraulic press to squeeze out the juice. The process had the advantage of preserving the contents of the yeast intact, splitting the cells without raising their temperature or exposing them to chemical substances. Having examined the results under

Floral pattern
The computer simulation (above) of a possible combination of enzyme structures, incorporating blue and yellow wheels, magenta florets and golden nanoparticles, is an example of one line of research undertaken by enzymologists.

Bioreactor
This maze of dials, pumps and tubes is a fermentation unit in which genetically modified bacteria are put to work producing proteins for industry.

MEET THE FAMILY

Enzymes belong to the protein family, a class of very large molecules that constitute an essential part of all living organisms, each formed of a chain of elementary constituents known as amino acids. Within cells, they are assembled by information provided from the genes, which produce an extraordinary variety through different combinations of just 20 acids. Enzymes are indispensable biological catalysts, but other proteins are equally essential to the organism's functioning, playing vital roles in connective tissue, hormones, antibodies and the operation of the nervous system.

a microscope to confirm that it contained no living cells, Buchner quickly went on to demonstrate that the press juice caused sugar to ferment, generating alcohol and carbon dioxide, just as living yeast cells would have done. He explained the fact by the presence of a protein that he christened 'zymase' in the extract. Proponents of the vital-force theory fiercely attacked the proposition, but he was able to refute all of their criticisms when he published his findings in 1897.

The birth of enzymology

Buchner had laid the foundations of the science of enzymology, and he was duly rewarded with the Nobel prize for chemistry in 1907. The biological implications of the new field were enormous, for it quickly turned out that enzymes were indispensable to the working of cells, concealed behind every fabrication, breakdown or transformation of molecules. Other Nobel prizes would be garnered by scientists following in Buchner's footsteps. Among them were Lwoff, Jacob and Monod, who were together awarded the 1965 prize for their work on the regulation of enzyme activity in genes, and Arber, Nathans and Smith, joint winners in 1978 for the discovery of so-called 'restriction enzymes' that can cut DNA sequences, making it possible to manipulate DNA for a variety of research purposes.

In a wider context, enzymology played a significant role in such major advances as the mapping of the genome and the identification of a range of illnesses linked to enzyme deficiency or malfunction. It has also turned out to have important industrial applications. These include not just washing powders that boast of 'dirt-eating' enzyme action or a role in the fermentation of foodstuffs stretching back over many decades, but also a variety of new uses that have only begun to be explored in the present century, the creation of genetically modified organisms among them. For scientists, enzymes still have an exciting future.

The tape recorder 1898

In 1898 Valdemar Poulsen invented the Telegraphone, precursor of the tape recorder. The Danish inventor reckoned that people making telephone calls would like to have some way of preserving conversations. Sound recording equipment had been available since 1877, when Thomas Edison introduced the phonograph. Poulsen took his inspiration from another American, Oberlin Smith, who had written an article suggesting the use of magnetic impressions to record sound. He replaced Edison's phonograph cylinder, coated in tinfoil, with one wrapped in steel wire. The two poles of an electromagnet rested against the wire, rotating down its length as the recording was made. A microphone transformed the sonic vibrations into electrical signals that were picked up by the magnet, leaving the wire magnetised to varying degrees as the sounds rose and fell. For playback, the microphone was replaced by a telephone receiver. The magnet was returned to the top of the wire, and as it slid back down again the recorded message could be heard.

Listening in *Early recording devices, like this one from 1907, recorded sound in the form of electrical signals that left a magnetic imprint on steel wire.*

Putting the tape in tape recording
In the 1930s German inventors substituted steel tape for the wire in Poulsen's machine. Plastic recording tape first appeared after 1945.

Poulsen's device proved a success when it was demonstrated at the 1900 Paris World's Fair, yet it had its drawbacks. The sound quality was poor and users had to put on headphones to hear the recordings. In addition, it took over a mile of wire to make a 15-minute recording. Even so, adapted versions of the mechanism later found their way into dictation machines The development of modern tape recorders had to wait on two discoveries made on the eve of the Second World War: valve amplifiers and modern recording tape, initially in the form of a strip of paper coated in iron oxide.

PRECURSOR
A hand-powered recording device employing wax paper rather than wire was patented by Alexander Graham Bell's Volta Laboratory in 1886. Listeners heard the recording through an ear-tube.

THE OLDEST RECORDED MESSAGE
Poulsen used the Telegraphone to record the voice of Franz-Josef I when the Austro-Hungarian emperor visited the Paris World's Fair in 1900. The recording is still preserved in Vienna's Technical Museum.

A powerful tool against pain and fever

Derived from salicylic acid extracted from willow bark, aspirin was first produced industrially by Bayer Laboratories at the turn of the 20th century. The popularity of this anti-inflammatory pain-killer has only grown in the ensuing decades.

Powder for pain
Bayer produced this box of soluble aspirin in about 1900.

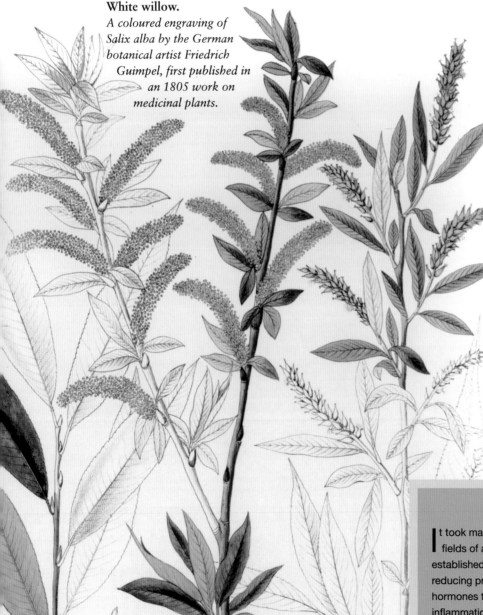

White willow.
A coloured engraving of Salix alba by the German botanical artist Friedrich Guimpel, first published in an 1805 work on medicinal plants.

On 25 April, 1763, the Reverend Edward Stone, a clergyman from Chipping Norton in Oxfordshire, presented a paper to the Royal Society in London on the use of willow bark to combat fever. Inspired by tales of the 'Jesuit bark', derived from the Peruvian chinchona tree and used by South American missionaries to treat malaria, he prepared a decoction of willow bark that he tested out experimentally on 50 people suffering from fevers. It proved a happy discovery – or, to be more correct, a rediscovery, since the efficacy of willow extract against inflammations had been described as early as 1550 BC in the ancient Egyptian Ebers Papyrus. Nevertheless, Reverend Stone's experiment marked the first time that the virtues of the bark had been demonstrated scientifically.

The first extracts

In 1826 a German chemist named Johann Andreas Buchner isolated the active ingredient in the bark, naming it salicin from the Latin *salix*, 'willow'. A couple of years later Henri Leroux, a French pharmacist, ground the

THE MYSTERIES OF ASPIRIN

It took many years for scientists to understand aspirin's multiple fields of action, but from the 1960s on John Vane and others established that it blocks an enzyme called cyclooxygenase, thereby reducing production of prostaglandins, substances related to hormones that play a part in the development of fevers, pain and inflammation. New properties are still being discovered, notably aspirin's ability to inhibit the development and recurrence of cancer of the colon as well as a possible role in preventing liver damage.

Quick! Aspro!

A 1964 advertisement aimed at the French market (above) recommends the virtues of Aspro against headaches. The brand name is one of a number under which aspirin-based medicines have been sold since Bayer lost exclusive global rights to the drug in 1919.

ALTERNATIVES TO WILLOW BARK

Willow bark may have been the first source of salicin, but compounds of the same family have since been traced in spiraea, beech, birch, liquorice and even strawberries. Nowadays synthetic substitutes can be made cheaply from easily available products such as phenol and acetic acid.

dried bark to make a powder that he then boiled in water, filtering the result and adding lead salts to precipitate the tannins. From 1.5kg of bark he obtained 30g of white salicin crystals, considerably improving the efficiency of the extraction process. Other scientists would clarify the properties of salicin and its derivatives, in particular salicylic acid. In 1853 Charles Frédéric Gerhardt, a chemist based in Strasbourg, synthesised acetylsalicylic acid by mixing salicylic acid with acetic acid, effectively inventing aspirin. But he died three years later, and his work was soon forgotten.

Queen of medicines

By the end of the 19th century the efficacy of salicin-based medicines was firmly established in the treatment of fever, inflammation and rheumatism. Yet their use was limited by their reputation for toxicity and causing stomach upsets, made worse by their unpleasant taste. In 1895 the German chemical firm Bayer, which specialised at the time in colorants, put researchers to work to find a solution to these

ANTI-COAGULANT

The Tsarevitch Alexis (1904–1918), only son and heir to Tsar Nicholas II of Russia, is said to have played a part in discovering the anti-clotting properties of aspirin; when the drug was prescribed for him, his haemophilia worsened.

problems. A team of a dozen or so chemists, including the young Felix Hoffmann, set to work alongside the pharmacologist Heinrich Dreiser and Dr Arthur Eichengrün. The story goes that Hoffman's father suffered from inflammatory rheumatism and that this fact may have inspired the son to rediscover acetylsalicylic acid. Whatever the truth of the tale, he managed to persuade Eichengrün to conduct experiments that revealed the drug's merits, which included analgesia.

In 1899 Bayer registered the trademark Aspirin, retaining exclusive rights until the end of the First World War. The drug quickly became one of the world's most popular medicaments, but was not available without a prescription until 1915.

In the 1900s researchers discovered that it also had anticoagulant properties. The idea of putting these to use to lessen the risk of heart attacks and strokes dates from the 1950s, but the drug's complex and enigmatic mechanisms initially led experts to treat the notion with caution. The use of low doses spread in the 1980s, largely thanks to work undertaken by John Vane, who won a Nobel prize in 1982 for research on aspirin's mode of action.

Protection for intellectual property in the modern era

As industry matured at the end of the 19th century, disputes between rival inventors became more common. The Paris Convention of 1883 represented a first attempt to establish a modern system for protecting intellectual property while at the same time stimulating innovation.

People have been devising new ways of doing things since the earliest times, and initially technological innovation spread freely from group to group and from generation to generation. As time passed, though, a need was felt to protect inventions by guaranteeing the people responsible for them exclusive rights to put them to use without fear of competition.

A statute passed in Venice in 1474 was the first to establish an exclusive right to exploit an invention. Yet such privileges soon proved open to abuse. In England there was mounting anger in the early 17th century against government powers to bestow monopolies controlling such staple commodities as paper, vinegar, salt and oil. In 1623 Parliament passed the Statute of Monopolies, abolishing all such arrangements except those granted to 'the first true Inventor or Inventors' of 'any manner of new Manufacture within this Realm' for a term not exceeding 21 years.

Rival systems
Antonio Meucci's 'speaking telegraph' (above) lost out to Alexander Graham Bell's application (left) in the race to patent the telephone.

Letter of intent
The inscription below comes from a patent application submitted to the French Academy of Sciences in April 1877 by Charles Cros, detailing a way of recording and playing back sound from a cylinder. Thomas Edison's phonograph was introduced less than a year later.

Limited duration

In the late 18th century American legislators inscribed in the US Constitution and in a law of 1790 exclusive moral rights to the exploitation of inventions. Every major nation followed suit. Intellectual protection gradually took the form it has today: inventors were rewarded for their efforts by the provision of sole rights to exploit their brainchild for a limited time in exchange for revealing the mechanisms involved, thereby stimulating further innovation. The length of the protected term varied and in 19th-century Britain it could be extended for a fee of up to £700.

The number of patents increased rapidly in the course of the 19th century, as did the frequency of disputes surrounding them, particularly if they cast their net so wide in describing the procedure or mechanism concerned as to obstruct the work of other innovators in the field. In 1854, for example, when photography was already attracting widespread interest, an American named James Cutting took out a patent on 'collodion positive' prints, also known as ambrotypes – images caught on sheets of glass using the wet plate collodion process. He subsequently claimed exclusive rights to the exploitation of collodion, even though it had already been used for years in other photographic processes.

Rival claims

There were challenges sparked by the borrowing or misappropriation of other people's inventions, the telephone being a prime example. Officially Alexander Graham Bell is accepted as the inventor of the first working telephone, but research has shown that a US-based Italian immigrant named Antonio Meucci produced a prototype of a 'speaking telegraph' in about 1854. Meucci was too poor to apply for a patent but instead took out a so-called 'patent caveat' – a legal

notice of intention to file for a patent at a later date. This had long expired by 14 February, 1876, when Bell was granted a patent on his invention, just a few hours before Elisha Gray filed another patent caveat on a similar device. After considering the rival claims, the US Patent Office awarded the patent to Bell. Yet Meucci's contribution to the telephone is undeniable, as the US House of Representatives recognised in 2002 when it passed a motion in honour of his achievement.

There were also cross-border disputes. In April 1886, a French physicist named Paul Héroult applied for a patent for a method of electrolysing an aluminium oxide solution to produce aluminium. An American engineer, Charles Hall, put in a similar application in the USA that July, repeating the exercise three years later in France, where his claim was rejected on the grounds that Héroult's had preceded it. Under American law, what mattered was that Hall had been the first to

apply the invention on US soil, and so his patent rights were upheld in 1892 even though Héroult's claim had primacy. Today the technique, which is still the standard industrial method of producing aluminium, is known jointly as the Hall-Héroult process.

The Paris Convention

Such controversies were frequent – so much so, in fact, that some inventors became nervous of showing off their discoveries at exhibitions and trade fairs in case they were stolen and copied. Gradually the idea of an international

WHAT CAN BE PATENTED?

For a patent to be assigned, an invention must be new, it must involve an 'inventive step that is not obvious to someone with knowledge and experience in the subject' and it must also have some industrial use. Works of art cannot be patented, and neither can scientific discoveries, although any practical applications may be eligible. In the UK, patents are granted for 20 years from the date of filing and are only valid within Britain. People sometimes jokingly refer to patenting 'a better mousetrap'; in fact James Henry Atkinson, a Leeds ironmonger, did just that in 1899 to protect the trap that became known as the 'Little Nipper', employing the familiar flat, spring-loaded design still in use today.

Early aviation patents
William S Henson took out a British patent for his Aerial Steam Carriage as early as 1843. In France, Clément Ader's first patent for a flying machine dates from April 1890; he built the Bat (above), his third prototype, in 1897. It was 1906 before the Wright brothers took out a US patent (top) on a machine that actually flew.

THE EUROPEAN PATENT OFFICE

In October 1973 the Munich Convention established the European Patent Organisation, with the European Patent Office (EPO) as its executive arm. The EPO cannot issue Europe-wide patents; instead, its job is to establish a common procedure for applications and the delivery of patents respected by all the signatories to the convention, currently totalling 35 separate states including all the EU nations. Successful applicants receive a bundle of separate national patents, each enforceable in the country in which it was issued.

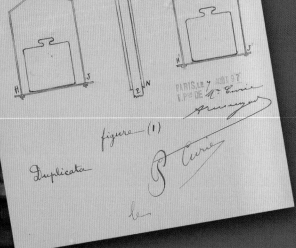

Signed and delivered
Pierre Curie's signature adorns a patent application dated 7 August, 1897, for a precision scale capable of measuring weights as fine as one hundredth of a milligram (above).

agreement gained ground. On 20 March, 1883, 14 nations signed the Paris Convention for the Protection of Industrial Property. When the treaty came into force, in July 1884, all the signatories agreed to grant the same degree of protection to foreign nationals as to their own citizens, not just in the patents themselves but also with regard to industrial designs and trademarks. Patent holders in one country were given a 'priority right' lasting six months (extended to a year in 1900) in which they can file an application for the same patent in the other signatory states.

Flaws in the system

Today patents are almost unanimously accepted as pillars of economic growth, yet inventors themselves sometimes complain about the cost and uncertainty of the system for obtaining them. A claim for patent protection might succeed in one country but not in another. European patents have to be translated into the official language of every country where they

An ill-starred inventor
Jean le Roy (left), a French photographer, invented an early cinematographic projector in 1895, but he never managed to patent his invention and thus missed out on any reward.

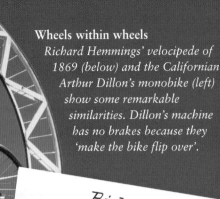

Wheels within wheels
Richard Hemmings' velocipede of 1869 (below) and the Californian Arthur Dillon's monobike (left) show some remarkable similarities. Dillon's machine has no brakes because they 'make the bike flip over'.

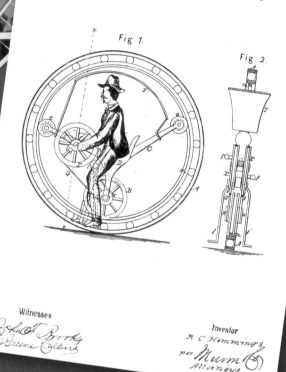

apply, and that can be expensive. Another criticism relates to patents on living organisms. Ethically and politically, these encourage the appropriation by private interests of common resources – particularly those of developing countries, where much of the world's biodiversity is to be found – and also of the personal biological data contained in biobanks.

Some observers argue that the current arrangement is a brake on innovation because it hinders competition, a view put forward by the American economists Eric Maskin and James Bessen in the year 2000 with particular reference to software patents. They maintained that innovators often need the freedom to play

around with preceding inventions, and that patents can inhibit this process.

The issue remains an open one. Other systems of intellectual protection have been proposed, from classic copyright arrangements to more restrictive versions of the current regime or to public bodies holding patent rights in common. Some brave souls have even given up entirely on the right to exclusivity, making software programmes and research findings common property freely available to all who want to use them.

PATENTS ON LIVING ORGANISMS

The first patents on a living organism were assigned to the French chemist Louis Pasteur, in 1865 in France and in 1873 in the USA, to protect a method of producing beer employing a special, purified yeast. Today it is not possible to take out a patent on a living organism itself, although 'technical instructions' detailing a method of improving some characteristic of the organism or of giving it new properties can be eligible. Following this principle patents have been granted on many gene sequences, including human ones, although these cannot be applied to naturally occurring genes.

Blood gives up its secrets

By identifying the principal blood groups, the Austrian physician Karl Landsteiner helped to make blood transfusions safe. Forty years later the same man was responsible for breakthrough work on the Rhesus factor and antigens; the Landsteiner–Wiener system still bears his name.

Memory aid
A diagram in note-pad format (bottom left) shows the compatibility between the four main blood groups, with speckling indicating clotting. Only Group O is compatible with all three others, which is why Group O individuals are sometimes called universal donors.

Identity test
Specific antigens are used to identify the blood groups to which different samples belong (below, centre). If the sample blood clots in the presence of an A antigen, for example, it is labelled Group A.

By the dawn of the 20th century doctors already had occasional recourse to blood transfusions, the first successful one having been performed in England in 1829. But while some turned out well, others continued to end in tragedy when the recipient, for apparently inexplicable reasons, rejected the blood given and died as a result. For the medical world these setbacks remained an enigma – one that was finally resolved thanks to the persistence and sagacity of a 32-year-old Austrian doctor and chemist, Karl Landsteiner.

Identifying blood groups

Born the son of a renowned Viennese journalist, Landsteiner became fascinated by the study of human blood while working as an assistant in the Pathological-Anatomical Institute at the University of Vienna. Between conducting autopsies he set himself the task of throwing light on the mechanisms responsible for blood coagulation (clotting).

One day Landsteiner took samples of his own blood and that of his fellow-workers and in each sample separated the serum, as the liquid part of the blood is called, from the red cells – the corpuscles responsible for carrying oxygen from the lungs to the body tissues. Then he tried out various combinations, mixing serum from one sample with cells from another. In some cases nothing at all happened, but in others the cells clotted. Landsteiner managed in this way to identify three types of blood, which he classified as A, B and C.

In Group A the surface of the red cells was covered with A antigens (the molecules that define the group's identity), while the serum clotted Group B red cells. Conversely, Group B serum clotted Group A cells. In the third group, which Landsteiner called C, the serum clotted both A and B cells, but its own cells reacted to neither of the serums of the other two groups. This group would later be renamed O, short for the German *ohne*, 'without', since the cells did not carry either the A or B antigen. Another, much rarer group, named AB, was subsequently identified in 1907 by Landsteiner's fellow-researchers while working with a larger group of volunteers.

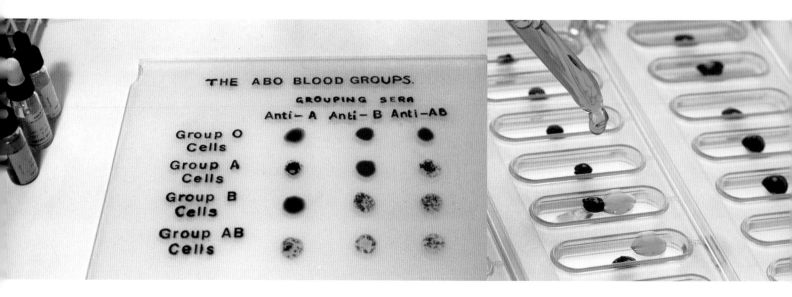

Close-up view
In this image of blood cells seen through an electron microscope, red blood cells are shown in red and platelets in brown. The white blood cells are artificially coloured here green, magenta and blue.

Centrifugal separation
To separate the different constituents of blood – red and white cells, for example, or plasma – according to their different densities, the blood is placed in a centrifuge machine (below) and spun at high speed.

A natural phenomenon

These simple, methodical experiments marked a date in the history of medicine. 'The serum of men in good health has a clotting effect not just on animal species but also on the blood corpuscles of different individual humans' Landsteiner noted in 1900. He soon became convinced that the clotting was a natural phenomenon rather than the consequence of some illness, as many doctors claimed at the time. It was a matter of a reaction between antigens and antibodies, with every individual developing specific antibodies as a defensive shield against antigens they did not have.

Landsteiner quickly envisaged some of the important implications the discovery of the ABO blood grouping would have, not just for transfusions but also for forensic medicine. In 1902 he explained to Vienna's Society for Forensic Medicine how analysing blood deposits at a crime scene could help identify the murderer. Yet the most dramatic impact was undoubtedly on blood transfusion.

By matching the blood group of donors to recipients it became possible to avoid most of the tragedies that had previously occurred in the transfer of blood between individuals. Yet one problem still remained to be resolved: the tendency of blood to coagulate as soon as it was taken from the donor. The gradual introduction of anticoagulants such as sodium

Landsteiner at work
The Austrian physician peers through a microscope in his New York laboratory. He moved with his family to the USA in 1923 to take up a post with the Rockefeller Institute, and eventually became an American citizen. He died of a heart attack in his laboratory in 1943, with a pipette still in his hand.

BLOOD GROUPS AND HEREDITY

The antigens of the ABO blood group system pass down the generations according to Mendel's laws of heredity, with each individual possessing one gene inherited from the mother and one from the father. A and B genes are dominant, while the O gene is recessive, which is to say that it only becomes apparent in the absence of dominant genes. The combination in a subject of Group A can be either AA or AO, just as a subject of Group B is BB or BO; someone in Group O is always OO and someone in Group AB is always AB. These properties were used in the course of the 20th century to determine paternity, but now more precise DNA tests are normally used.

citrate made possible the huge increase in the number of transfusions administered in the years during and after the First World War.

A revolution in surgery

The prospect of risk-free transfusions had another important consequence for surgeons. In future they were able to undertake operations on the heart, lung and blood vessels that had previously been impossible because of the risk of massive haemorrhages.

Saving lives
American soldiers in the Second World War unload containers of blood intended for on-the-spot transfusions in the field.

Landsteiner emigrated to the USA in 1923, accepting an invitation to carry on his research at the Rockefeller Institute in New York. Four years later he and his fellow-researcher Philip Levine identified new blood groups labelled M, N and P. He was rewarded for his efforts with the 1930 Nobel prize for medicine. Even

then, his exceptional career – which also included important discoveries on the agents responsible for polio and syphilis – was far from over. In 1940, at the age of 71 and officially retired, he was still hard at work in his laboratory, collaborating with Alexander Wiener to develop a serum, derived by immunising rabbits with blood cells from rhesus macaque monkeys, that reacted with about 85 per cent of human red blood cells.

By the time of his death in 1943, blood cells had given up many of their secrets. Yet even so, the Landsteiner did not live long enough to learn that white blood cells also carry antigens. Over the course of the ensuing decades the HLA system which plays a significant role in transfusions – the letters stand for 'human leukocyte antigen' – was gradually deciphered.

THE FIRST BLOOD BANKS

In the early days blood transfusions were performed on a case-by-case basis; a compatible donor needed to be found ready to hand. The situation became more flexible in the years between 1914 and 1918 with the discovery of the first anticoagulants, and also with the introduction of refrigeration as a way of storing blood. These advances enabled Oswald Hope Robertson, a British-born doctor, to set up 'blood depots' for casualties in France in the First World War.

By the mid 1930s, Soviet Russia had a nationwide system that involved transporting blood supplies to some 60 provinicial centres. The term 'blood bank' itself was coined in 1937 when the USA got its first permanent facility, established at the Cook County Hospital in Chicago. In the course of the Second World War doctors learned how to separate blood in order to extract easily useable derivatives like plasma and albumen. Later it became possible to 'wash' blood in order to remove unwanted clotting agents.

Meccano 1900

The inventor of Meccano was Frank Hornby, in his thirties at the time and working as a bookkeeper for a Liverpool meat importer. He came up with the idea while trying to make a toy crane with his two young sons. The concept was simple enough: the kits provided the equipment needed to create models by assembling tin strips (later they were made of steel) perforated with equally spaced holes, with the aid of screws, nuts and washers. The idea of interchangeable parts was central; bolted together in different positions to make separate designs, they could easily be disassembled to make new ones.

Encouraged by the kit's success within his immediate family circle, Hornby patented the concept on 9 January, 1901, under the name 'Mechanics Made Easy', shortened to Meccano in 1907. More than a century after its invention Meccano remains a familiar name worldwide, not just for children but also among adult model-makers, engineering enthusiasts and collectors.

Constructive fun
As the 1920s advertisement below suggests, Meccano has helped generations of children to learn the pleasure of making things. The detail (right) is from a giant model in the town hall at Calais in France, now home to a major Meccano manufacturing plant.

AN EMPIRE BUILT ON TOYS

Hornby was responsible for more than just Meccano, significant though that was. In 1934 he introduced die-cast model cars and trucks that became famous as Dinky Toys. Four years later his firm started selling Hornby Dublo train sets, originally featuring clockwork as well as electric locomotives, which quickly became Britain's leading model railway kit.

The paper clip 1900

The Byzantines were the first to use paper clips, made out of bronze, to fasten imperial documents. An American, Samuel B Fay, patented triangular bent-wire fasteners in 1867, but they were not produced in significant numbers until the turn of the century. The familiar looped-wire paper clip came into use at about the same time; the earliest known advertisement for the 'Gem', as it was called, has been traced to 1894. Oddly, the design was never patented, although the trademark 'Gem' was, in 1904. Humble though they may seem, paperclips took on symbolic significance during the Nazi occupation of Norway. When the authorities banned the wearing of national symbols, Norwegian patriots took to wearing them in their lapels as a symbol of solidarity in the face of foreign rule.

An early 20th-century city risen from the ashes

The USA's second city after New York, Chicago was rebuilt within a few years of being ravaged by a terrible fire in 1871. Skyscrapers rose up to crowd the skyline, the stockyards and the business district hummed with activity, and the Mafia began to weave its web.

Surveying the damage
Chicago residents examine the desolation left in the wake of the Great Fire of 1871.

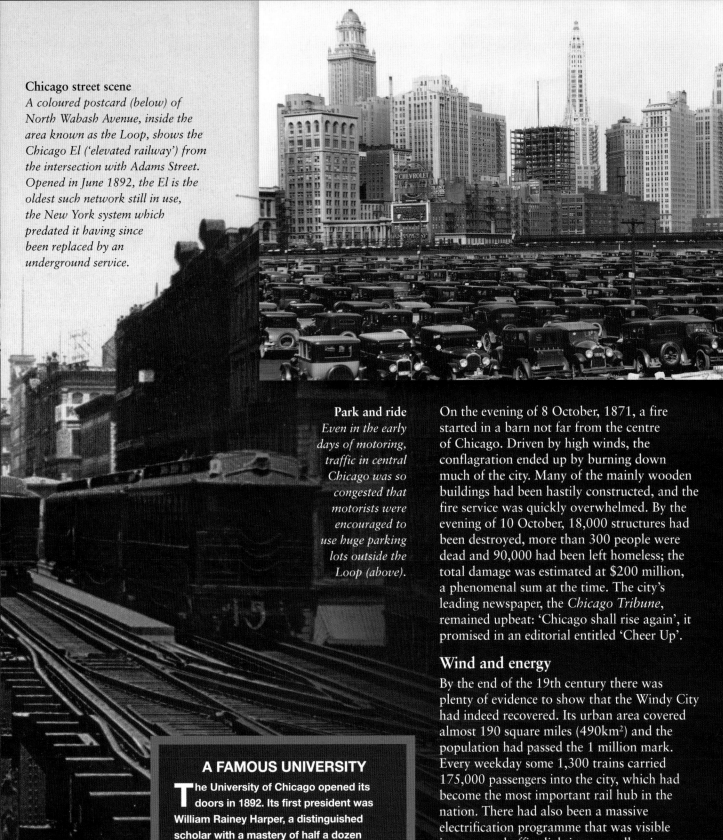

Chicago street scene
A coloured postcard (below) of North Wabash Avenue, inside the area known as the Loop, shows the Chicago El ('elevated railway') from the intersection with Adams Street. Opened in June 1892, the El is the oldest such network still in use, the New York system which predated it having since been replaced by an underground service.

Park and ride
Even in the early days of motoring, traffic in central Chicago was so congested that motorists were encouraged to use huge parking lots outside the Loop (above).

On the evening of 8 October, 1871, a fire started in a barn not far from the centre of Chicago. Driven by high winds, the conflagration ended up by burning down much of the city. Many of the mainly wooden buildings had been hastily constructed, and the fire service was quickly overwhelmed. By the evening of 10 October, 18,000 structures had been destroyed, more than 300 people were dead and 90,000 had been left homeless; the total damage was estimated at $200 million, a phenomenal sum at the time. The city's leading newspaper, the *Chicago Tribune*, remained upbeat: 'Chicago shall rise again', it promised in an editorial entitled 'Cheer Up'.

Wind and energy

By the end of the 19th century there was plenty of evidence to show that the Windy City had indeed recovered. Its urban area covered almost 190 square miles (490km²) and the population had passed the 1 million mark. Every weekday some 1,300 trains carried 175,000 passengers into the city, which had become the most important rail hub in the nation. There had also been a massive electrification programme that was visible in street and office lighting, as well as in telephone exchanges and tramways.

In its rise from the ashes, the 'City of Big Shoulders', as it was sometimes known, had turned to some of the world's finest architects, who lined its main arteries with metal-framed skyscrapers equipped with new high-speed lifts and central heating. These revolutionary structures concentrated business activity in a high-density central area of banks, luxury hotels and wide-windowed offices that provided workspaces for many thousands of

A FAMOUS UNIVERSITY

The University of Chicago opened its doors in 1892. Its first president was William Rainey Harper, a distinguished scholar with a mastery of half a dozen ancient languages. The university boasted the USA's first sociology department and was also the first institution of its kind to offer adult evening classes. It soon attracted an elite group of professors; the head of the physics department was Albert Michelson, who in 1907 became the first American to win a Nobel prize in the sciences when his work on the speed of light was recognised with the award for physics.

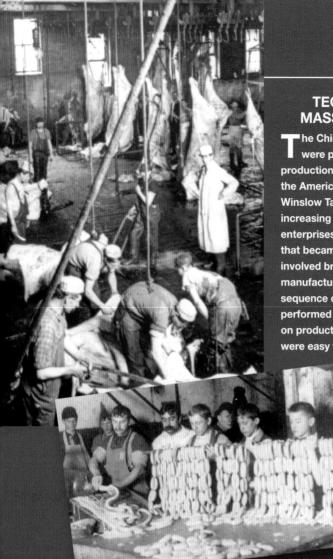

Making meat
Some of the meat from the beasts slaughtered in the stock yards (top left) found its way into sausages produced at factory premises nearby – this photograph (left) was taken in 1893. The last abattoir in Chicago closed in 1971 along with the Union Stock Yards.

employees. Real-estate prices in the Loop at the city's heart rose sevenfold between 1880 and 1890, symbolising Chicago's economic resurgence. Ringed by the El, as the Elevated overground metro system was already known, and crisscrossed by cablecars and a growing number of automobiles, the Loop hummed with life from dawn to dusk. A host of white-collar workers, the men in derby hats and the women in black skirts and white blouses, hurried to their workplaces from the suburbs each morning, while wealthy individuals with time on their hands set out to window-shop in the big stores on Madison and State streets.

Of cattle and men

Millions of tonnes of wheat and maize passed though Chicago each year, making it the world's largest cereal marketplace. Its role as a transportation hub also made it the redistribution centre for timber for all the western states. It was best known, though, as the focus of the US meat industry, which was concentrated in the Packingtown area (now known as New City) to the south of the Loop. At its height, between 25,000 and

30,000 people were employed in the stock yards and abattoirs, earning Chicago, an almost unavoidable stopping-off point between the Mid Western grasslands and the East Coast population centres, ironic nicknames like Porkopolis and Hog Butcher of the World.

Every year some 13 million animals arrived in Chicago, where they were herded into vast pens while their owners haggled with the packers who controlled the meat market. The beasts were then slaughtered, bled and hung from meat-hooks on moving production lines that carried them to workers who each had a specific butchering task to perform. Every carcass passed through dozens of hands before the meat was stored in refrigerated warehouses.

The workers, most of them recent immigrants, laboured in dreadful conditions for wretched wages, going home each evening to makeshift accommodation in cheap tenement blocks. Strikes and rioting broke out in May 1886 in support of the demand for an eight-hour working day. A bomb thrown during a demonstration in Haymarket Square killed seven policemen. Three anarchists were hung for the deed, sowing terror in the workers' ranks. By 1890 almost three-quarters of Chicago's inhabitants were foreign-born, grouped by nationality in a number of ethnic neighbourhoods. The influx of young, often single men provided plenty of custom for the saloons, gambling houses and brothels in the Levee, the city's red-light district. The owners of such establishments maintained shady links with some of the city's leading politicians.

City of culture

But Chicago was much more than the nation's sausage capital or the 'Western Gomorrah' where the Mafia was already establishing its roots, later to blossom under Al Capone in the 1920s. Charles Lawrence Hutchinson, a rich banker who wanted to make the city the 'Florence of the American West', founded the Art Institute on Michigan Avenue, where Chicagoans could admire works by the old

AN INSPIRATION TO HENRY FORD

In his autobiography *My Life and Work*, published in 1922, the industrialist Henry Ford revealed that the idea of using a moving production line to assemble automobiles 'came in a general way from the overhead trolley that the Chicago packers use in dressing beef'. He reasoned that if such methods could be used to handle up to 1,000 animal carcasses an hour, they could be adapted to construct cars on the same principle.

masters for nothing on Sundays, when admission charges were dropped. Close by, the Public Library opened its doors to everyone, whether to borrow books or read them on the premises. A bevy of newspapers, among them the *Chicago Daily News*, *Chicago Record*, *Chicago Sun-Times* and *Chicago Herald*, as well as the *Tribune*, competed fiercely for readers with the aid of the latest advances in Linotype machines and printing presses.

The World's Fair of 1893, which celebrated the 400th anniversary of Christopher Columbus's arrival in the Americas, put the reborn Chicago firmly on the map. Between May and October that year, more than 27 million visitors came from around the globe to gather on the banks of Lake Michigan, where a gigantic 'White City', lit by 65,600 incandescent bulbs and 3,426 arc lamps, had sprung up to glorify technological progress. Today the metropolis that gave birth to Walt Disney and Ernest Hemingway, as well as to *Playboy* magazine, Wrigley's chewing gum and the McDonalds restaurant chain, is more than ever the archetypal American city.

A TEMPLE TO THE GLORY OF ELECTRICITY

The Chicago World's Fair of 1893 was the first to devote an exhibition hall entirely to electricity. Visitors could admire different applications of the new form of energy, including Bell telephones (the first call between Chicago and Boston was made on 7 February that year), Edison's phonograph, which could record an entire opera on a series of cylinders, and Elisha Gray's Telautograph, a precursor of the fax machine that electronically copied handwriting.

World-class view
Spectators enjoy a panoramic view of the Chicago World's Fair of 1893. More than 200 structures were erected for the event, among them the first giant Ferris wheel.

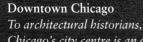

Downtown Chicago
To architectural historians, Chicago's city centre is an open-air museum of high-rise building styles.

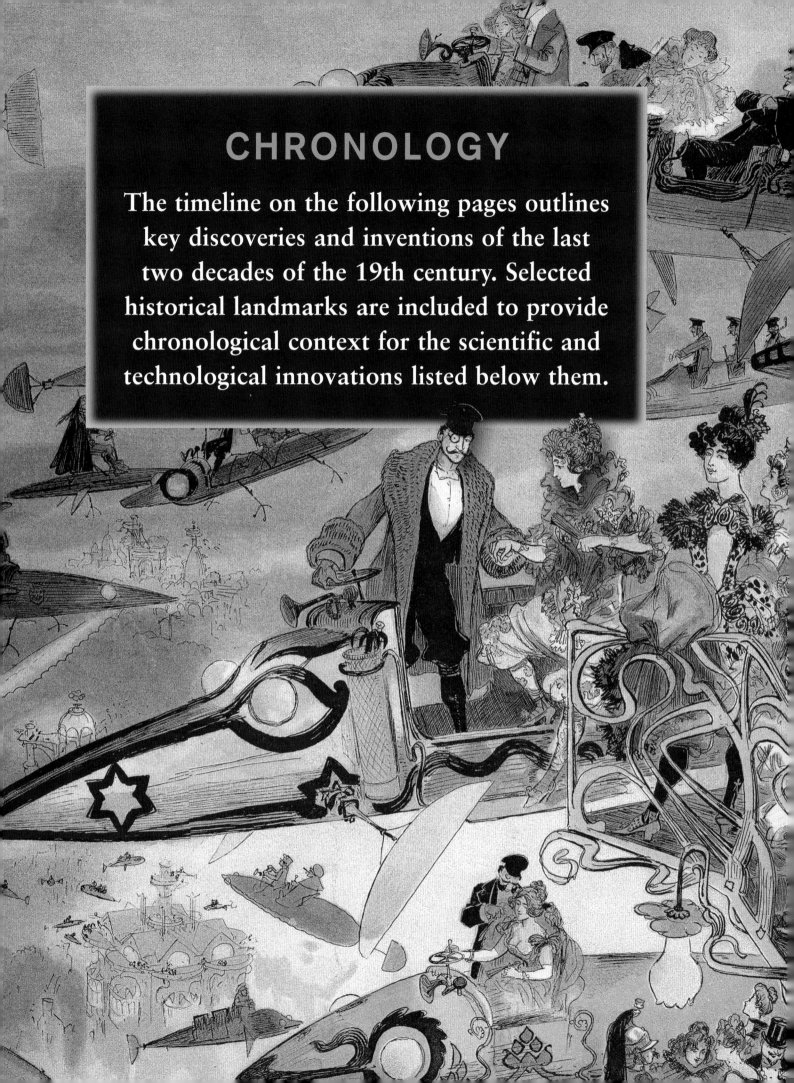

CHRONOLOGY

The timeline on the following pages outlines key discoveries and inventions of the last two decades of the 19th century. Selected historical landmarks are included to provide chronological context for the scientific and technological innovations listed below them.

1873

- Proclamation of the 1st Spanish Republic (1873)
- Restoration of the Bourbons (1874)
- Britain annexes Fiji (1874)

- Remington & Sons start producing the first commercial typewriter

- Levi Strauss & Co begin making denim overalls reinforced with rivets

1875

- Abdulhamid II becomes absolute ruler of the Ottoman Empire (1876)
- Custer's last stand at Little Bighorn (1876)
- Queen Victoria proclaimed Empress of India (1877)
- War breaks out between Russia and Turkey (1877)

- Oscar Hertwig first observes fertilisation through a microscope, unveiling the secrets of the reproduction of species

- Alexander Graham Bell files for a patent on the telephone; the units of measurement for sound volume (bels and decibels) will be named in his honour

- Charles Cros and Thomas Edison conceive of the phonograph almost simultaneously, but Edison is the first to build a working model; a fecund and multifaceted inventor, Edison will also become renowned as a pioneer of modern business methods, linking technological and industrial progress

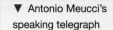

◀ Alexander Graham Bell's patent application for the telephone

▼ Antonio Meucci's speaking telegraph

◀ Werner von Siemens' *Elektromote*, the first trolley bus

1878

- Treaty of Berlin ends the Russo-Turkish War (1878)
- The Second Anglo-Afghan War breaks out (1878)
- The Anglo-Zulu War ends Zulu independence (1879)
- The War of the Pacific breaks out in which Chile fights an alliance of Bolivia and Peru (1879)

- Paris hosts its third World's Fair (1878)

- Thomas Edison and Joseph Swan independently devise the first practical light bulbs

- Constantin Fahlberg and Ira Remsen discover saccharin, the world's first artificial sweetener

- James Ritty, an Ohio saloon-keeper, invents the cash register

- Lester Pelton introduces the Pelton Wheel, a new and efficient form of water turbine

1880

- Britain recognises an independent Transvaal following the First Boer War (1881)
- Alexander III becomes Tsar of Russia (1881)
- The Triple Alliance links Germany, Austro-Hungary and Italy (1882)
- British forces bombard Alexandria and occupy Cairo in Egypt (1882)

- The world's first electrical power station is installed in Godalming, Surrey

- Britain's John Dixon Gibbs and Lucien Gaulard from France invent a high-voltage electrical transformer

- The French inventor Clément Ader invents the electrophone, introducing the principle of stereophonic sound

- In Berlin, Werner von Siemens demonstrates his *Elektromote*, a precursor of the trolley bus

- Étienne Jules Marey develops the chronophotographic camera, an ancestor of the cine-camera, which enables him to photograph living creatures in motion; a prolific and visionary innovator, Marey is also a pioneer of aeronautics and the inventor of the sphygmograph, a device for recording blood pressure on paper

- Frederick Winslow Taylor develops the doctrines of scientific management that will become known as Taylorism, seeking to break down work into its component parts in order to increase productivity; he will set out his ideas in *The Principles of Scientific Management*, published in 1911

▲ Ladies listening in to Clément Ader's electrophone

▲ The Midvale Steel Company in Nicetown, Pennsylvania, a factory run on Taylorist, mass production lines

▶ Power station on Niagara Falls

1883

- Bismarck introduces sickness insurance for workers in Germany (1883)
- The Berlin Conference speeds up the colonial Scramble for Africa (1884)
- Germany annexes South-West Africa (Namibia, 1884)
- General Gordon occupies Khartoum in the Sudan (1884)

- The American Lewis Edison Waterman patents the capillary-feed fountain pen

- The Paris Convention for the Protection of Industrial Property marks a first step toward the introduction of an international patent system

- The city of Paris introduces a regular door-to-door garbage collection

- Germany's Otto Schott discovers borosilicate glass, the first truly efficient industrial glass

- In Chicago William Le Baron Jenney designs the metal-framed Home Insurance Building, often claimed to be the world's first skyscraper

1885

- Gordon is killed as the Mahdi's forces retake Khartoum (1885)
- Belgium's King Leopold II establishes personal control over the Congo (1885)
- Germany annexes Tanzania as the protectorate of German East Africa (1885)
- Serbia defeated in war against Bulgaria at Slivnitza (1885)

- An Atlanta pharmacist, John Styth Pemberton, introduces Pemberton's French Wine Cola, soon to be renamed Coca-Cola

- In Germany Wilhelm Maybach puts an engine on a bicycle frame to create the first motorbike

- German-born American citizen Ottmar Mergenthaler invents Linotype – the first hot-metal typesetting system using a machine that produces entire lines of copy

- Clemens Winkler discovers the element germanium

- The mechanisms of nitrogen fixation by plants are discovered

- Nikola Tesla, a Serbian-born American citizen, founds the Tesla Electric Light & Manufacturing Company to exploit applications of his invention of the alternating-current (AC) induction motor; Tesla will be the principal champion of AC electricity in the 'war of the currents' with Thomas Edison, who favours the DC alternative

► Collecting household waste in Paris

▼ Waterman pen nibs

► Triple-lens microscope by Carl Zeiss

1887

- Queen Victoria celebrates her Golden Jubilee (1887)
- Slavery abolished in Brazil (1888)
- Chile annexes Easter Island (1888)
- Kaiser Wilhelm II succeeds to the German throne (1888)

• Talbot Lanston patents the Monotype system, a rival to Linotype

• Ludwik Zamenhof promotes the use of an international language of his own invention: Esperanto

• The Hungarian David Gestetner invents the duplicating machine

• John Boyd Dunlop develops the pneumatic tyre

• George Eastman puts the first Kodak camera on the market

• Theophilus Van Kannel invents the revolving door

• Heinrich Hertz demonstrates the existence of electromagnetic waves beyond the visible spectrum

• Thomas Edison assigns two of his employees, Harold P Brown and Arthur E Kennelly, the task of developing a method of execution using the alternating current championed by his commercial rival; the work will lead to the electric chair

1889

- The Meiji Constitution gives Japan a constitutional monarchy (1889)
- Crown Prince Rudolf of Austro-Hungary dies, an apparent suicide, at the Mayerling hunting lodge (1889)

• Édouard Michelin develops the removeable tyre

• George Eastman introduces transparent celluloid film

• The world's first jukebox is installed in San Francisco's Palais Royale Saloon

• Santiago Ramón y Cajal reveals the structure of nerve cells, laying the foundations for modern neuroscience

• In France Herminie Cadolle introduces a two-part undergarment that will evolve into the modern bra

• Albert Gay and Robert Hammond patent the electric kettle

• Conceived as the centrepiece of Paris's 1889 World's Fair, the Eiffel Tower is the culminating achievement of the French engineer Gustave Eiffel (1832-1923), previously responsible for groundbreaking designs for bridges and railway viaducts as well as the inner framework of the Statue of Liberty

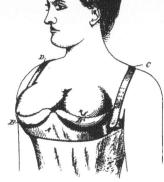

▲ The 'Breast Supporter', an early bra patented by Marie Tucek

▲ Stairwell inside the Statue of Liberty

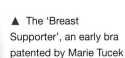

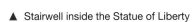

▶ Stand publicising the merits of Esperanto

1891

EVENTS

• The International Peace Bureau is founded, with headquarters in Bern, Switzerland (1891)
• The Franco-Russian Alliance is formed to counter the Triple Alliance (1892)

INVENTIONS

• William Dickson, an employee of Thomas Edison, invents the Kinetoscope

• The first electric oven to reach the market goes on sale in the USA

• Whitcomb Judson patents the 'clasp locker', a precursor of the zip fastener

• James Dewar invents (but fails to patent) the Dewar flask, the precursor of the Thermos flask

• Jesse W Reno invents the escalator

• Ronald Ross, a Scottish doctor serving in India, identifies the mosquito responsible for spreading malaria

• Dmitri Ivanovski, a Russian botanist, first describes a virus

• Rudolf Diesel patents the self-igniting compression engine that will come to bear his name

1893

• The Corinth Canal opens in Greece (1893)
• Women in New Zealand are granted the vote (1893)
• The Dreyfus Affair gets under way in France (1894)
• The First Sino-Japanese War breaks out over Korea (1894)

• France's Léon Appert invents shatterproof glass

• Caleb Bradham invents Pepsi-Cola

• Carl Benz introduces the Victoria, his first four-wheel automobile

• Chicago plays host to the 1893 World's Fair

• The German firm Hildebrand & Wölfmuller manufacture the world's first production motorcycle

• The world's first motor race is staged in 1894 between Paris and Rouen in France

• A Geneva clockmaker, Casimir Sivan, produces the first speaking clock

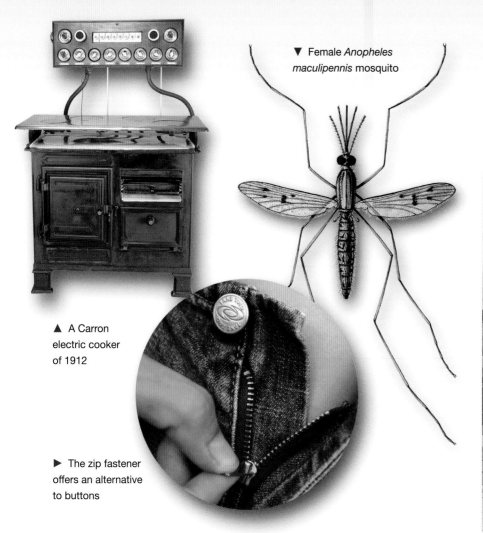

▼ Female *Anopheles maculipennis* mosquito

▲ A Carron electric cooker of 1912

▶ The zip fastener offers an alternative to buttons

1895

- The Treaty of Shimonoseki ends the first Sino-Japanese War in Japan's favour (1895)
- Ethiopian forces defeat an Italian expeditionary force in the first Italo-Ethiopian War (1896)
- Theodor Herzl, the 'father of Zionism', publishes *The Jewish State* (1896)

- The Lumière brothers organise the world's first public cinema show in Paris

- Wilhelm Röntgen, a German physics professor, discovers x-rays

- King Camp Gillette invents the safety razor with reuseable blades

- John William Cockerill, a British army surgeon, has the idea of placing a small rear-view mirror above a car's windscreen

- Manchester-born Joseph John Thomson discovers electrons

- The first Olympic Games of modern times are staged in 1896 in Athens on the initiative of a French educationalist, Baron Pierre de Coubertin

- The French physicist Henri Becquerel discovers natural radioactivity

- Georges Méliès establishes the world's first cinema production company; he introduces the use of special effects in films and establishes the first film studio – in the garden of his house in a suburb of Paris

1897

- The Greco-Turkish War ends in Greece sueing for peace (1897)
- The USA annexes Hawaii (1898)
- Victory in the Spanish-American War gives the USA control over Cuba, Puerto Rico, Guam and the Philippines (1898)

- Taxi meters, invented in Germany by Wilhelm Bruhn, first come into service

- Liverpool's School of Tropical Medicine opens its doors – the first institution of its kind in the world

- The German bacteriologist Friedrich Loeffler discovers the virus responsible for foot-and-mouth disease

- The husband-and-wife team of Pierre and Marie Curie discover the radioactive elements polonium and radium; after Pierre's death in 1906 Marie continues to confirm her position as one of the world's leading scientists, winning a second Nobel prize for her work on radioactivity

- Eduard Buchner, a German chemist, helps to establish the science of enzymology

- Valdemar Poulsen from Denmark invents the Telegraphone, a precursor of the tape recorder

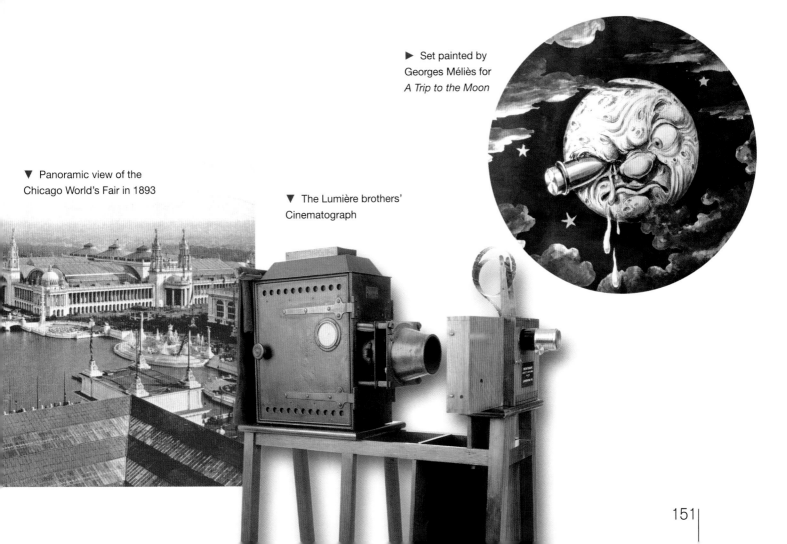

► Set painted by Georges Méliès for *A Trip to the Moon*

▼ Panoramic view of the Chicago World's Fair in 1893

▼ The Lumière brothers' Cinematograph

1899

EVENTS

• The First Hague Peace Conference leads to the signing of the Hague Convention (1899)
• The Second Boer War begins in southern Africa (1899)
• The Boxer Rebellion breaks out in Peking (Beijing) in China (1900)

INVENTIONS

• Working for the German chemical firm Bayer, Felix Hoffman and his associates develop aspirin by rediscovering the analgesic and anti-inflammatory properties of acetylsalicylic acid, first synthesised almost 50 years earlier

• French physicist Paul Villard discovers gamma rays

• Paris hosts the 1900 World's Fair

• In Paris in 1900 women participate in the Olympic Games for the first time, although only in tennis and golf

• Frank Hornby devises the model construction kit that will come to be known as Meccano

• Karl Landsteiner identifies the principal blood groups

• The Canadian inventor Reginald Fessenden makes a first pioneering radio transmission

• Clementines are first grown in Algeria

• Raoul Grimoin-Sanson introduces Cinéorama, an early experiment in 360° multi-screen viewing, at the Paris World's Fair

1901

• The Commonweath of Australia is established (1901)
• Nigeria becomes a British protectorate (1901)
• Death of Queen Victoria; she is succeeded by Edward VII (1901)

• Guglielmo Marconi makes the first trans-Atlantic radio transmission

• The first teddy bears are produced in Germany and the USA, where they take their name from President Theodore Roosevelt

• Working at the Edison Laboratory, Miller Reese Hutchison develops the Acousticon, the first electric hearing aid

• Dutch physiologist Willem Einthoven develops the first electrocardiogram

• Working in Birmingham, Frederick Lanchester patents the first disc brake for cars

▼ White willow (*Salix alba*) – Aspirin was first produced from salicin extracted from its bark

▲ An early Meccano kit

▶ The 'Gem' paper clip

1903

- The Entente Cordiale is established between Britain and France (1903)
- War breaks out between Russia and Japan (1904)
- The Herero Revolt begins in German Southwest Africa (1904)

- Pierre and Marie Curie and Henri Becquerel share the Nobel prize for physics for their work on radioactivity

- Mary Anderson patents a design for automobile windscreen wipers in the USA

- German chemists Joseph von Mering and Emil Fischer synthesise the first commercial barbiturate, marketed as Veronal

- Mikhail Tsvet, a Russian botanist, develops chromatography as a method for separating plant pigments

- In the USA the Wright brothers make the first controlled heavier-than-air flights in their pioneer aeroplanes, the Wright Flyer I and II

- Elizabeth Magie Philips takes out a US patent on *The Landlord's Game*, an antecedent of *Monopoly*

- Ira Washington Rubel develops a method of offset printing for use on paper

1905

- Strikes and demonstrations spread across Russia (1905)
- Norway breaks away from union with Sweden (1905)
- The Algeciras Conference puts Morocco under international control (1906)

- Death of Jules Verne, author of an impressive body of work often said to represent the birth of science fiction; in the 54 novels he published under the general heading of *Extraordinary Voyages*, he foresaw the exploration of the Moon and of the ocean floor, the launching of satellites and many other subsequent developments

- Albert Einstein publishes his theory of special relativity

- English physiologists Ernest Starling and William Bayliss first identify hormones

- Joseph John Thomson is awarded the Nobel prize for physics for the discovery of the electron

▼ Pierre and Marie Curie

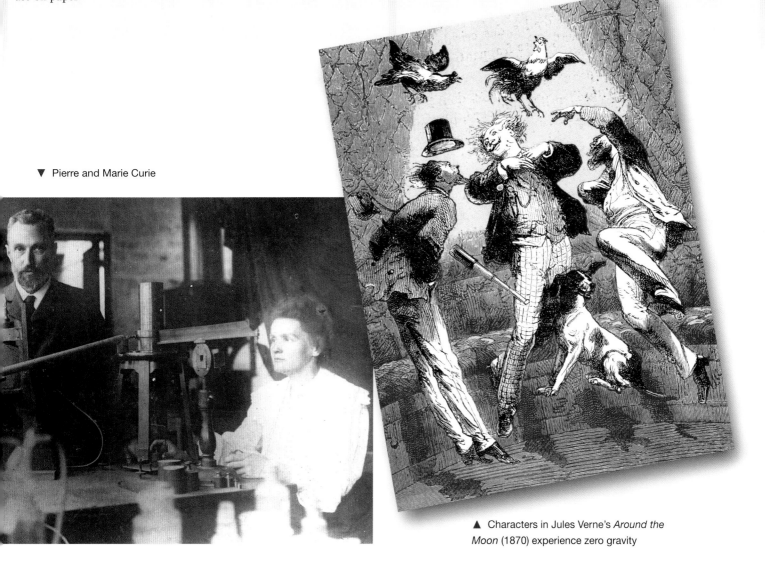

▲ Characters in Jules Verne's *Around the Moon* (1870) experience zero gravity

Index

Page numbers in *italics* refer to captions.

Picture credits

Front cover: main image, the Palace of Electricity at the World's Fair in Paris (1900), Leemage/ Photo Josse/Musée Carnavalet, Paris/H Hoffbauer. **Inset**: an early French car in the 'landaulet' body style with an open driver's seat, RMN/ P Segrette/Musée de l'Armée, Paris.
Spine: the iconic Coke bottle, Delagarde J P.
Back cover: an early tape recording machine, Cosmos/SSPL/Science Museum, London.
Page 2, left to right, top row: Corbis/DK Limited; Leemage/Majno; Cosmos/SSPL/Science Museum; 2nd row: Delagarde J P; Corbis/ Underwood & Underwood; Cosmos/SSPL/Science Museum; 3rd row: AKG Images/SPL; Collection NLC; US Library of Congress/Motion Picture/ Dickson/Edison; bottom row: DR; Bridgeman Art Library/The Stapleton Collection; Roger-Viollet.
Pages 4/5: view of Chicago, Tim Boyle/Getty Images; 6t: Picture Desk/Dagli Orti/National Photographic Company, Chicago/L Platt Winfrey; 6/7: Bridgeman Art Library/Giraudon; 7tl: AKG Images/Sotheby's; 7tr: Leemage/MP; 7br: Corbis/DK Limited; 8tl: The Picture Desk/Dagli Orti/The Art Archive; 8bl: AKG Images; 8/9t: BNF/Bibliothèque Nationale de France, Paris/ 7857713; 8/9b: © Musée des Arts et Métiers-CNAM, Paris/Photo J-C Wetzel; 9c: Corbis/ F T Harmon; 9br: Cosmos/ SSPL/NMeM/Royal Photographic Society; 10tl: Cosmos/SSPL/ Science Museum; 10tr: Leemage/Aisa/Faculty of Medicine, Madrid; 10bl: Bridgeman Art Library/Bibliothèque des Arts Décoratifs, Paris/Archives Charmet; 10/11: Corbis/ Bettmann/Lt Lubbe; 11bl: Kharbine-Tapabor/ G Roux; 11r: Corbis/Bettmann; 12cl: Leemage/ Selva; 12tr: Library of Congress; 12b: Leemage/ Fototeca; 12/13: US Library of Congress/Motion Pictures/Dickson/Edison; 13t: Corbis/Hulton Deutsch Collection; 13b: AKG Images; 14tl: Leemage/North Wind Pictures; 14c: Leemage/ Selva; 14br: Leemage/Palais Longchamp, Marseille/J Bernard; 14/15: AKG Images; 15tr: Cosmos/SSPL/Science Museum; 15b: AKG Images; 15cr: Longines/DR; 16l: Cosmos/ SSPL/Science Museum; 16/17t: Leemage/Photo Josse/Bibliothèque de l'Institut Curie, Paris; 16d: Cosmos/SPL; 17l & 17tr: Cosmos/SSPL/ Science Museum; 17br: Corbis/ Bettmann/ Underwood & Underwood; 18/19: Corbis/Bettman; 20t: Leemage/North Wind Pictures; 20b: Library of Congress/Prints and Photographs Division; 21l: Leemage/North Wind Pictures; 21r: Cosmos/ SSPL/Science Museum Library; 22tl: Corbis/ Bettmann; 22br: AKG Images; 23: Jupiter Images/ Dynamic Graphic; 24t: Leemage/Bianchetti; 24b: Collection NLC; 25bl: Leemage/Gusman/ Musée de l'École, Carcassonne; 25c: Leemage/ Bianchetti; 25tr: Leemage/Majno; 26bl: Leemage/Bianchetti; 26: Jupiter Images/Photo Objects; 27t: Corbis/Y Forestier; 27b: REA/Denis; 28/29t: Bridgeman Art Library/Giraudon; 28b: Leemage/Majno; 29r: Leemage/Bianchetti; 29bc: AKG Images/Sotheby's; 30b: Cosmos/ SPL/S Volker; 30t: Corbis/Sagel & Kranefeld; 31t: Image courtesy of the Advertising Archives; 31c: Delagarde J P; 31r: Corbis/Bettmann; 32t: US Library of Congress/Prints and Photographs Division; 32b: Corbis/DK Limited; 33tl: Corbis/Bettmann; 33tr: Cosmos/SPL/CERN; 34cl: Corbis/Bettmann; 34t: Leemage/MP; 35c: PROD DB/Lucasfilm/DR/Star Wars, 1977; 35b: Cosmos/SPL/V Habbick Vision; 36: Corbis/ Bettmann; 37t: AKG Images; 37c & 38t: Cosmos/

SSPL/Science Museum; 38b: Leemage/ Bianchetti; 39t: Cosmos/M Fairbanks; 39b: Corbis/EPA/V Janninck; 40bl & br: The Picture Desk/Dagli Orti/The Art Archive; 40/41: US Library of Congress/Prints and Photographs Division; 41tr: Collection NLC; 44b: BNF/ Bibliothèque Nationale de France, Paris/ 7844094; 42t: Cosmos/SPL/Dr J Burgess; 42b: Cosmos/ SPL/Gustoimages; 43t: AKG Images/Ullstein Bild; 43b: Corbis/Underwood & Underwood; 44t: CNAM/Musée des Arts et Métiers; 44b: © Musée des Arts et Métiers-CNAM, Paris/Photo J-C Wetzel; 45t: College de France Archives; 45c: BNF/Bibliothèque Nationale de France, Paris/7857713; 45b: Cosmos/SSPL/ NMeM; 46bl: Bridgeman Art Library; 46tr: Corbis/ F T Harmon; 47t: US Library of Congress/ Prints and Photographs Division/Underwood & Underwood; 47b: Cosmos/Aurora/J Azel; 48: Cosmos/R Crandall; 49bl: AKG Images; 49tr: Cosmos/SSPL/NMeM; Royal Photographic Society; 50l: Historical Society of Western Pennsylvania, Library and Archives Division, Pittsburgh; 50/51t: RMN/BPK/G Buxenstein Company; 51c: PROD DB/UFA/DR/Metropolis, Fritz Lang, 1926; 51br: Corbis/Bettmann/Lt Lubbe; 52l: Leemage/Aisa/Faculty of Medicine, Madrid; 52r: Cosmos/SPL/A Pasieka; 53: Cosmos/SPL; 54b: Cosmos/SPL/Laguna Design; 54t: Bridgeman Art Library/Bourneville & Regnard; 55bl: Bridgeman Art Library/Archives Charmet/Charles Lemmel/ADAGP, Paris 2010; 55tr: Collection NLC; 56t & b: Cosmos/SSPL/ Science Museum; 57: Corbis/Bettmann; 58: The Picture Desk/Dagli Orti/The Art Archive/Culver Pictures; 59bl: Cosmos/N Benn; 59c: Corbis/ Lake County Museum; 59r: Corbis/A Woolfitt; 60t: Corbis/Bettmann/Ch C Ebbets; 60b: Corbis/ Bettmann; 61tl: Corbis/Ch E Rotkin; 61r: Corbis/R Sulgan/architects I M Pei & C Pelli; 62tl: Bridgeman Art Library/The Stapleton Collection/S Walery; 62b: Leemage/Lee; 63tl: Bridgeman Art Library/Bibliothèque des Arts Décoratifs, Paris/Archives Charmet; 63br: Corbis/M L Stephenson; 64 background: AKG Images/D Bellon; 64tr: AKG Images; 64b (2): Bridgeman Art Library/Giraudon/Musée Carnavalet, Paris; 65: Leemage/Bianchetti; 66t: Corbis/DPA/M Hanschke; 66c: AKG Images/ SPL; 66br: REA/L Hanning; 67: Leemage/Selva; 68b: AKG Images/M Zapf; 68t: Bridgeman Art Library/The Stapleton Collection; 69t: Cosmos/ SPL/V Steiger; 69c: Cosmos/SPL/S Stammers; 69br: Cosmos/SPL/Dr Tony Brain; 70: Leemage/ MI/Musée Jules-Verne, Nantes, Gustav Wertheimer; 71l: Leemage/Selva; 71r: Leemage/Selva; 72l: Kharbine-Tapabor/G Roux; 72/73: Corbis/Bettmann; 73tl: Cosmos/SPL/NASA; 73tr: Kharbine-Tapabor/E Bayard; 73b: AKG Images/Bibliothèque, Amiens Métropole/ E Bayard; 74l: Leemage/Fototeca; 74t: Leemage/Gusman/A Robida; 74b: AKG Images; 75bl: Roger-Viollet; 75tr: Leemage/Costa; 76c: Corbis/Visual Unlimited; 76/77: Cosmos/ SPL/UCT/Dr L Stannard; 77b: Cosmos/Eye of Science; 78c: Corbis/Bettmann; 78tr: Corbis/ Hulton Deutsch Collection; 79t: Cosmos/SPL/TEK Image; 79b: Cosmos/ SPL/R Harris; 80bg: Collection NLC; 80br: Leemage/Selva; 81tl: AKG Images; 81tr: REA/Sinopix/C Galloway; 82c: Deutsches Historisches Museum, Berlin/ L Schimdt; 82/83: Leemage/Photo Josse/Musée Carnavalet, Paris/H Hoffbauer; 83tr: Picture Desk/Dagli Orti/National Photographic Company, Chicago/L Platt Winfrey; 83br: Roger-Viollet; 84/85 & 86/87: photo F Kleinefenn/*la Fée*

Électricité, détails, Raoul DUFY © ADAGP, Paris 2009; 88bl: RMN/BPK; 88tr: US Library of Congress; 89t: RMN/Musée de la Voiture, Compiègne/J G Berizzi; 89b: Picture Desk/Dagli Orti/Bodleian Library, Oxford; 90l: Leemage/ Photo Josse; 90tr: Leemage/Fototeca; 91t: Corbis/SYGMA/R Reuter; 91br: RMN/Musée de la Voiture, Compiègne/J P Lagiewski; 92l: RMN/Musée de l'Armée, Paris/P Segrette; 92c: Cosmos/SSPL/Science Museum; 92/93: Leemage/Costa; 93t: CNAM/Musée des Arts et Métiers; 93tr: Cosmos/SSPL/Science Museum; 93bl: RMN/Musée de la Voiture, Compiègne/J P Lagiewski; 93br: Reuters/Toru Hanai; 94tl: US Library of Congress/Motion Picture/Dickson/Edison; 94bl: Bridgeman Art Library/Giraudon/NMPTF, Bradford; 94/95: AKG Images; 95cl: AKG Images; 96t: RMN/BPK; 96c: Leemage/Delius; 97tr: Collection Archives Françaises du Film du CNC/Films Scientia/Eclair; 97b: Corbis/Museum of History and Industry; 98t: AKG Images/SPL; 98b: Corbis/J Springer Collection; 99t: AKG Images; 99b: The Picture Desk/KOBAL Collection/Gaumont; 100: Bridgeman Art Library/Lawrence Steigrad Fine Arts, New York; 101 (2): Corbis/Hulton Deutsch Collection; 102t: Corbis/Bettman; 102/103c: Leemage/ Selva; 103t: Corbis/Hulton Deutsch Collection; 103b: Leemage/Selva; 104: Cosmos/SSPL; 105t: Leemage/North Wind Pictures; 105b: AKG Images; 106 (2): RMN/BPK; 107bl: Cosmos/ SPL/M Fermariello; 107t: Cosmos/SPL/NASA; 108tl & tr: Roger-Viollet; 108br: Corbis/PhotoAlto/ L Hamels; 109t: Jupiter Images/Comstock; 109b: Cosmos/Aurora/L Howlett; 110bl: DR; 110c: Leemage/Palais Longchamp, Marseille/ J Bernard; 110/111: RMN/ Musée d'Orsay/ Ch A Lhermitte/H Lewandowski; 111t: Corbis/ Alinari Archives/Edizioni Brogi; 111cr (2): DR; 111br: Image courtesy of the Advertising Archives; 112l: The Picture Desk/Dagli Orti/Culver Pictures; 112/113: AKG Images; 113br: RMN/Médiathèque du Patrimoine/Ministère de la Culture; 114: Cosmos/SSPL/Science Museum; 115l: Corbis/ Bettmann; 115br: Cosmos/SPL/L Berkeley Laboratory; 116bl: Cosmos/SPL; 116tc: Cosmos/ SPL/CTR Wilson; 117cl: Cosmos/SPL/ Dr F Espenak; 117tr: Cosmos/SPL/Pasieka; 117br: Cosmos/SPL/CERN; 118b: AKG Images; 118c: AKG Images/Ullstein Bild; 118br: Longines/DR; 118/119tr: AKG Images/J Hios; 119: Leemage/Heritage Images; 120b: Cosmos/ SPL/C Powell, P Fowler & D Perkins; 120t: Cosmos/SSPL/Science Museum; 121: Cosmos/ SPL/M Garlick; 122b: Cosmos/SSPL/Science Museum; 122c: Cosmos/SPL/A & H Frieder Michler; 122/123: Leemage/Photo Josse/ Bibliothèque de l'Institut Curie, Paris; 123b: Leemage/Imagestate; 124tr: Leemage/Selva; 124r: Cosmos/SPL; 125tl: Musée Curie (collection ACJC)/Institut Curie; 125cr: Leemage/Selva; 125br: Leemage/Heritage Images/Science Museum; 126: Corbis/Visual Unlimited; 127tl: Cosmos/SPL/Laboratory of Molecular Biology, Cambridge/Dr A Lesk; 127d: Brookhaven National Laboratory/J Hainfeld, L Quian, M Hu; 128: Cosmos/Rosenfeld Images Ltd; 129b: Cosmos/SSPL/Science Museum; 129t: US Library of Congress/Prints and Photographs Division; 130b: AKG Images/F Guimpel; 130t: Cosmos/ SSPL/Science Museum; 131tl: Corbis/Firefly Production; 131tr: Bridgeman Art Library/ Giraudon/R Savignac/ADAGP, Paris 2010; 132bl: Leemage/Gusman/Académie des Sciences, Paris; 132c: AKG Images/SPL; 132tr: Leemage/Costa; 133c: Bridgeman Art Library/

Giraudon; 133tr: Cosmos/SSPL/NASA; 134b: Corbis/Bettmann/Underwood & Underwood; 134tr: BNF/Bibliothèque Nationale de France, Paris/22003219; 135tl: Cosmos/E Sander; 135cr: Corbis/Bettmann; 136bl: Cosmos/SSPL/ Science Museum; 136br: Cosmos/SPL/ M Fermariello; 137bl: Cosmos/SPL/Ch Priest; 136/137: Corbis/ Visual Unlimited; 137tr: Corbis/Bettmann; 138: Corbis/Bettmann; 139bl: Collection NLC; 139cr: Image courtesy of the Advertising Archives ; 139tr: GLOBEPIX/ F Huiban; 140l: Corbis; 140/141: US Library of Congress/Prints and Photographs Division/ Washington; 141tr: Corbis/Bettmann/Underwood & Underwood; 142tl: US Library of Congress/Prints and Photographs Division/Washington; 142cl: US Library of Congress/Prints and Photographs Division/Underwood & Underwood/Washington; 143tl: Corbis; 143b: HEMIS FR/P Frilet; 144/145: Leemage/Gusman/A Robida; 146l: AKG Images/SPL; 146c: Leemage/Costa; 146b: Collection NLC; 146/147: RMN/BPK/G Buxenstein Company; 147r: Leemage/Bianchetti; 147b: US Library of Congress/Prints and Photographs Division; 148c: Leemage/Bianchetti; 148bl & br: Leemage/Majno; 148/149: Corbis/ M L Stephenson; 149b: AKG Images/Ullstein Bild; 149r: Collection NLC; 150l: Cosmos/SSPL/ Science Museum; 150b: Corbis/DPA/M Hanschke; 150c: Leemage/Selva; 150/151: Corbis; 151b: Bridgeman Art Library/Giraudon/NMPTF, Bradford; 151r: Corbis/Bettman; 152bl: AKG Images/ F Guimpel; 152cr: Image courtesy of the Advertising Archives; 152br: Collection NLC; 152/153: Musée Curie (collection ACJC)/Institut Curie; 153br: Kharbine-Tapabor/E Bayard.

THE ADVENTURE OF DISCOVERIES AND INVENTIONS
Into the Electric Age – 1880 to 1900
Published in 2010 in the United Kingdom by Vivat Direct Limited
(t/a Reader's Digest), 157 Edgware Road, London W2 2HR

Into the Electric Age – 1880 to 1900 is owned and under licence from
The Reader's Digest Association, Inc. All rights reserved.

Adapted from *L'Ère de l'Électricité*, part of a series entitled L'ÉPOPÉE DES
DÉCOUVERTES ET DES INVENTIONS, created in France by BOOKMAKER and
first published by Sélection du Reader's Digest, Paris, in 2010.

Translated from French by Tony Allan

PROJECT TEAM
Series editor Christine Noble
Art editor Julie Bennett
Designer Martin Bennett
Consultant Ruth Binney
Proofreader Ron Pankhurst
Indexer Marie Lorimer

Colour origination FMG, London
Printed and bound in China

VIVAT DIRECT
Editorial director Julian Browne
Art director Anne-Marie Bulat
Managing editor Nina Hathway
Picture resource manager Sarah Stewart-Richardson
Technical account manager Dean Russell
Product production manager Claudette Bramble
Production controller Sandra Fuller

We are committed both to the quality of our products and the service we provide to our
customers. We value your comments, so please feel free to contact us on 08705 113366
or via our website at **www.readersdigest.co.uk**

If you have any comments or suggestions about the content of our books, you can
email us at **gbeditorial@readersdigest.co.uk**

CONCEPT CODE: FR0104/IC/S
BOOK CODE: 642-008 UP0000-1
ISBN: 978-0-276-44520-0
ORACLE CODE: 356400008H.00.24